Eclipse Almanac
2021 To 2030

Ten Years of Solar and Lunar Eclipses

BLACK AND WHITE EDITION

Fred Espenak

Edition 1.0
October 2020

Eclipse Almanac 2021 To 2030 – Black and White Edition

Ten Years of Solar and Lunar Eclipses

Astropixels Publishing
P.O. Box 16197
Portal, AZ 85632

www.astropixels.com/pubs

This book may be ordered at: *www.astropixels.com/pubs/EclipseAlmanac.html*

More about solar and lunar eclipses from 2021 to 2030 can be found at EclipseWise:

www.astropixels.com/news/EclipseAlmanac.html

Astropixels Publication Number: AP025

First Edition (Version 1.0a)

ISBN 978-1-941983-25-6

Printed in the United States of America

Front Cover: Drawing of the Sun and Moon is attributed to Albrecht Dürer, *Nuremberg Chronicle* (Hartmann Schedel, 1493).

Back Cover: Portrait of Fred Espenak (Copyright ©2018 by Fred Espenak).

Preface

The *Eclipse Almanac* contains maps and diagrams of every solar and lunar eclipse over a ten-year period. This permits the reader to look ahead and easily determine when and where each of these events will be seen. Particular details about each eclipse are included as well as a 25-year table looking further into the future.

Section 1 covers solar eclipses, while Section 2 is devoted to lunar eclipses. Brief explanations are given for the different types of eclipses, and descriptions of the visual appearance of each one is included. Section 3 tabulates the date and time of the Moon's phases over a decade. New Moon and Full Moon phases that coincide with solar and lunar eclipses, respectively, are identified.

The *Eclipse Almanac* series consists of five volumes. Each one covering a single decade as follows:

1) Eclipse Almanac 2021 to 2030   4) Eclipse Almanac 2051 to 2060
2) Eclipse Almanac 2031 to 2040   5) Eclipse Almanac 2061 to 2070
3) Eclipse Almanac 2041 to 2050

Each volume is available in a color edition as well as a more economical black and white edition.

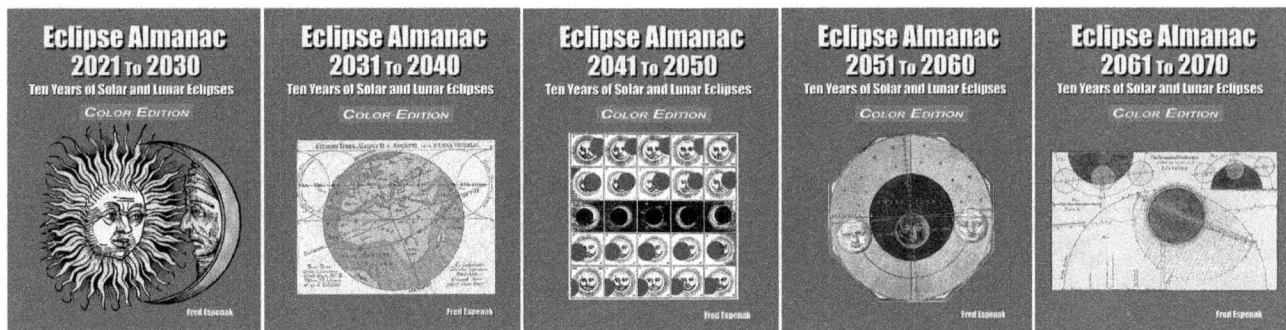

All times listed in the *Eclipse Almanac* are in Universal Time (UT1). This is the modern replacement of Greenwich Mean Time, which was based on mean solar time from Greenwich, England. In comparison, Universal Time is based on Earth's rotation with respect to distant quasars.

For North Americans, the conversion from UT1 to local time is as follows:

Atlantic Standard Time (AST)	= UT1 - 4 hours
Eastern Standard Time (EST)	= UT1 - 5 hours
Central Standard Time (CST)	= UT1 - 6 hours
Mountain Standard Time (MST)	= UT1 - 7 hours
Pacific Standard Time (PST)	= UT1 - 8 hours

If Daylight Saving Time is in effect in the time zone, you must ADD one hour to the standard time.

The conversion of Universal Time to other time zones is easily found on the Internet.

For more information about solar and lunar eclipses from 2021 to 2070, visit *EclipseWise.com* at:

www.astropixels.com/news/EclipseAlmanac.html

Table of Contents

Central Solar Eclipses from 2021 to 2030

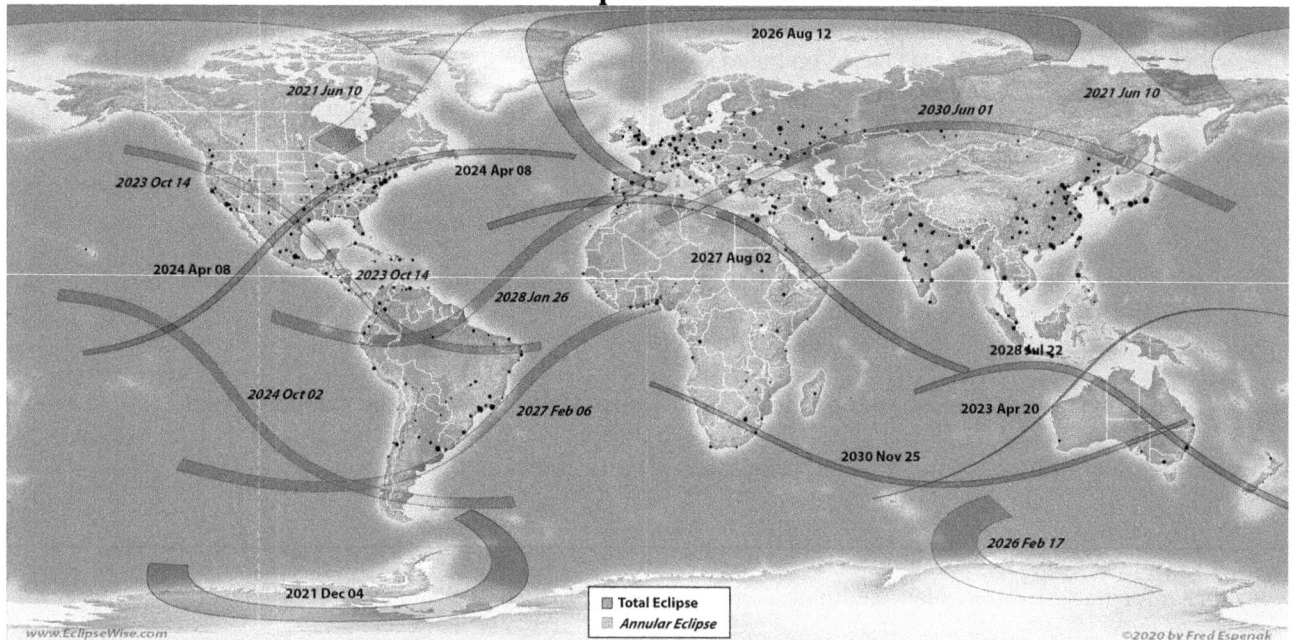

The path of every total, annular, and hybrid solar eclipse from 2021 to 2030 is plotted on a world map. Major cities are plotted as black dots, scaled by population size. (©2020 F. Espenak)

Photo 1–1 The "Diamond Ring Effect" is seen just before totality begins. Total Solar Eclipse of 2018 Jul 03. ©2018 F. Espenak

Section 1: Solar Eclipses

Introduction

The Moon orbits Earth once every 29.5306 days with respect to the Sun. Over the course of its orbit, the Moon's changing position relative to the Sun results in its familiar phases: New Moon > First Quarter > Full Moon > Last Quarter > New Moon. The New Moon phase is the only one not visible because the illuminated side of the Moon then points away from Earth.

The orbit of the Moon is tilted about 5.1° to Earth's orbit around the Sun. The points where the two orbits appear to cross are called the nodes. When the New Moon occurs near one of these nodes, the Moon's shadow falls on a portion of Earth and a solar eclipse is visible from that region.

The Moon's shadow is composed of three cone-shaped components. The outer or penumbral shadow is a zone where the Sun's rays are partially blocked. Nested within the penumbra is the umbral shadow — a region where direct rays from the Sun are completely blocked. The conical umbra tapers to a point beyond which extends an expanding cone called the antumbra. From within this third shadow, the Moon appears smaller than the Sun and is seen in silhouette against the solar disk.

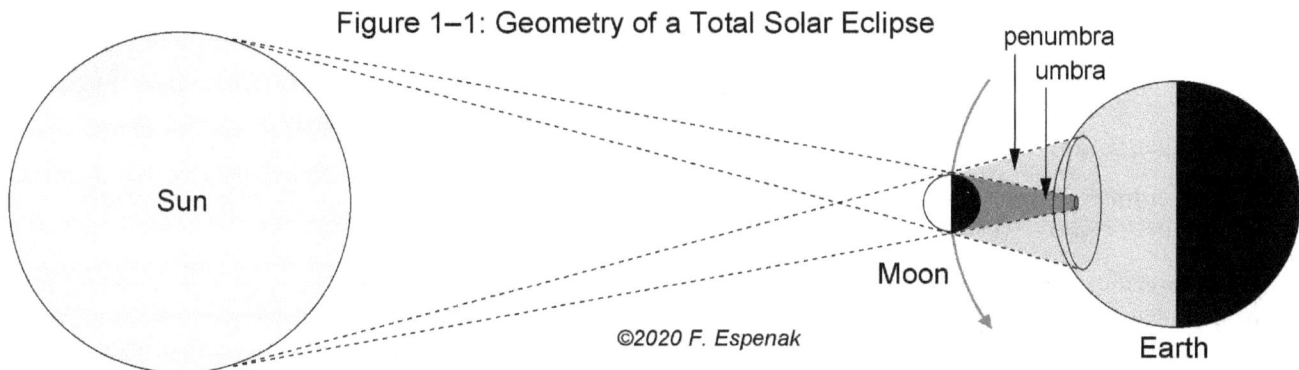

Figure 1–1 illustrates the geometry of a total solar eclipse. A partial eclipse is visible from within the large penumbral shadow, while the total eclipse is only seen from the much smaller umbral shadow.

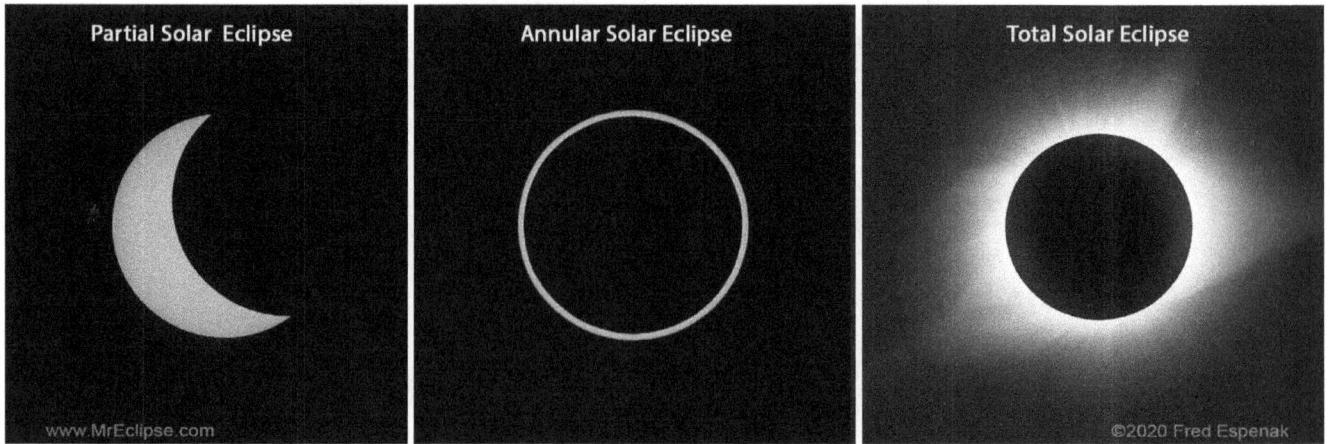

Photo 1–2 Examples of a partial, annular, and total solar eclipse. ©2020 F. Espenak

Types of Solar Eclipses

There are four types of solar eclipses:

1. **Total Solar Eclipse** — The Moon's penumbral and umbral shadows traverse Earth. The Moon's antumbral shadow extends beyond Earth's surface. The Moon appears larger than the Sun and completely covers the solar disk when viewed from within the umbral shadow. A partial eclipse is seen within the penumbral shadow (Figure 1–1).
2. **Annular Solar Eclipse** — The Moon's penumbral and antumbral shadows traverse Earth. The Moon's umbral shadow completely misses Earth. The Moon's disk appears smaller than the Sun so a bright ring of the Sun's disk surrounds the Moon when viewed from within the antumbral shadow. A partial eclipse is seen within the penumbral shadow (Figure 1–2).
4. **Hybrid Solar Eclipse** — The Moon's penumbral, umbral and antumbral shadows all traverse parts of Earth. The curvature of Earth's surface brings some regions into the umbra and others into the antumbra. The eclipse is total within the umbra and annular within the antumbra. A partial eclipse is seen within the penumbral shadow. Hybrid eclipses are also known as annular-total eclipses.
4. **Partial Solar Eclipse** — The Moon's penumbral shadow traverses Earth while the umbral and antumbral shadows miss Earth. A portion of the Sun's disk is obscured from within the penumbra.

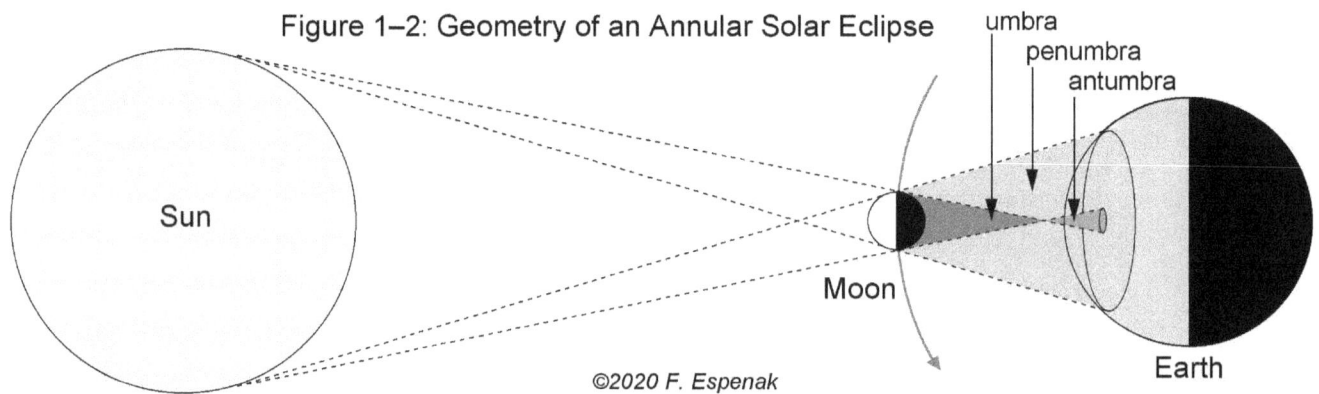

Figure 1–2 illustrates the geometry of an annular solar eclipse. A partial eclipse is visible from within the large penumbral shadow, while the annular eclipse is confined to the much smaller antumbral shadow.

Annular, total and hybrid eclipses are sometimes referred to as *central* eclipses[1]. However, on rare occasions it is possible to have an annular or total eclipse that is non-central. This occurs in Earth's polar regions when only the edge of the umbral or antumbral falls on Earth while the shadow axis misses the planet entirely. The eclipse of 2043 April 09 is a non-central total solar eclipse.

---

[1] A central eclipse is one in which the central axis of the Moon's umbral/antumbral shadow intersects with Earth's surface.

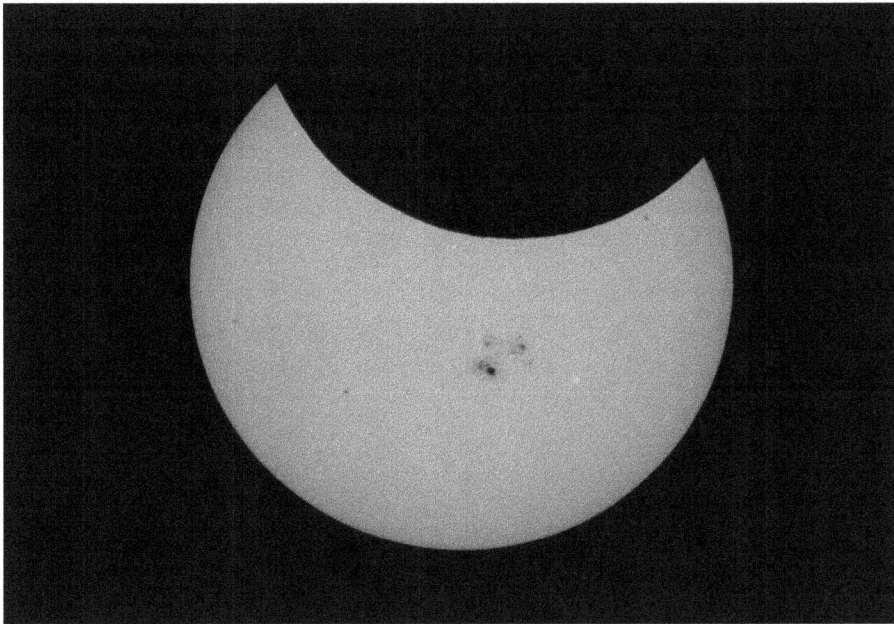

Photo 1–3 Partial solar eclipse of 2014 Oct 23. ©2014 F. Espenak

Visual Appearance of Partial Solar Eclipses

When the Moon's penumbral shadow strikes Earth, a partial eclipse of the Sun is visible from that region. The apparent motion of the Moon with respect to the Sun is gradual — the partial phases can two hours or more. During this time, the Moon's dark limb slowly creeps across the Sun's disk.

Partial eclipses cannot be viewed with the unprotected eye because the Sun is still extremely bright. Special techniques are needed to safely view the eclipse (for example, special eclipse filters or glasses). Even when a partial eclipse reaches its maximum phase, the sky and landscape remain bright. But careful inspection of dappled sunlight beneath a leafy tree will reveal multiple images of the eclipse. The gaps between the tree leaves act like pinhole cameras and project images of the crescent Sun onto the ground.

The Moon's penumbral shadow is typically 4200 to 4500 miles (6700 to 7300 kilometers) in diameter and can cover a significant fraction of the daytime hemisphere of Earth. Consequently, partial eclipses are visible from large geographic regions as the penumbra sweeps across Earth's surface.

Photo 1–4 shows various phases of the Annular solar eclipse of 2005 Oct 03. ©2005 F. Espenak

Visual Appearance of Annular Solar Eclipses

During an annular eclipse, the Moon's penumbral and antumbral shadows sweep across Earth. Compared to the penumbra, the antumbra is much smaller and has a maximum diameter of 232 miles (374 kilometers). Because of this, the antumbra covers a tiny fraction of Earth's surface.

A partial eclipse is visible within the penumbral shadow, but only observers located in the much narrower track of the antumbra will see an annular eclipse. For this reason, the antumbra's trajectory across Earth is called the path of annularity.

All annular eclipses begin with a series of partial phases lasting an hour or more. At the peak of the eclipse, the Moon's disk can be seen in complete silhouette against the Sun. The remaining solar photosphere appears as an intensely bright ring of light surrounding the Moon. The annular phase can last a maximum of 12 ½ minutes but is more typically 3 to 5 minutes in length. After annularity, another series of partial phases occur as the Moon gradually uncovers the Sun.

Special precautions must be used to watch an annular eclipse (just like partial eclipses). Even during the annular phase, the Sun is dangerously bright and cannot be viewed without a solar filter. The landscape and sky remain bright throughout the eclipse, giving little indication of the celestial event in progress.

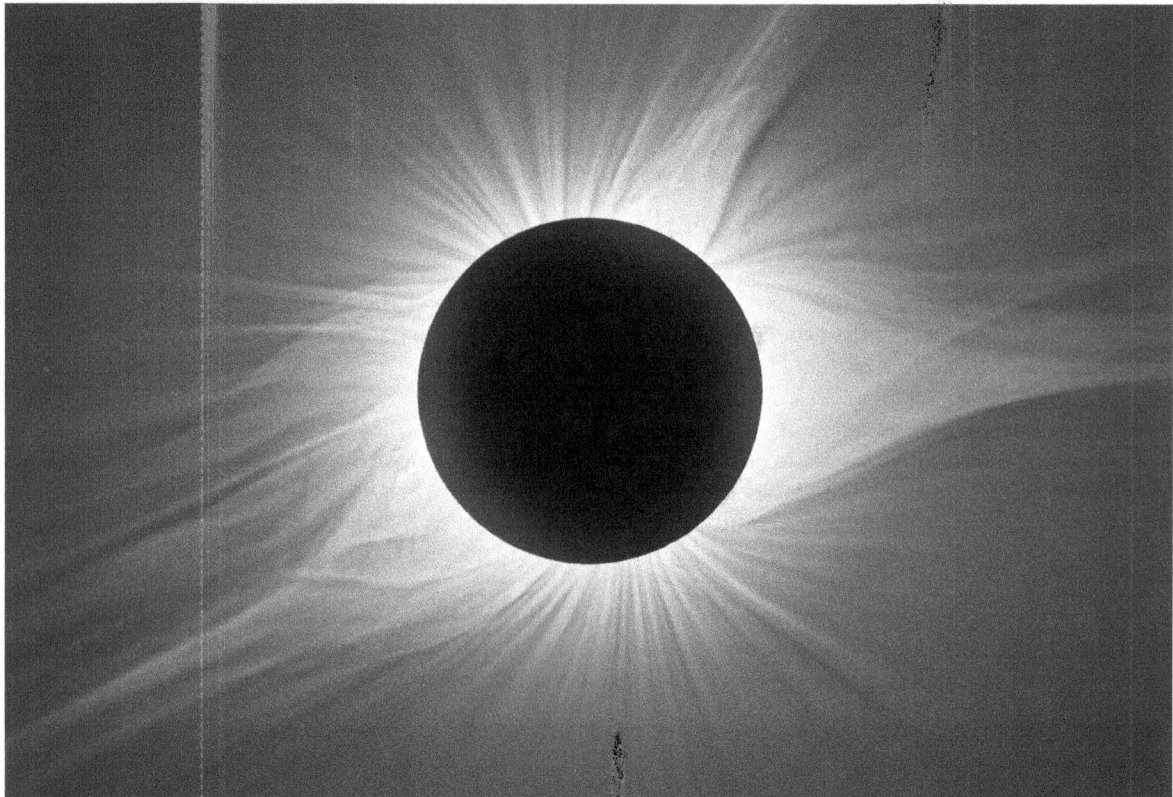

Photo 1–5 The Sun's corona is only visible to the naked eye for a few minutes during a total eclipse of the Sun. This is an HDR composite image of the total solar eclipse of August 21, 2017 from Casper, Wyoming. © 2017 F. Espenak, www.MrEclipse.com

Visual Appearance of Total Solar Eclipses

During a total eclipse, the Moon's penumbral and umbral shadows fall upon Earth. The umbra has a maximum diameter of 170 miles (273 kilometers). The narrow track traced out as the umbra sweeps across Earth's surface is called the path of totality. The Sun is completely obscured by the Moon from within this zone. The total phase can last up to 7 ½ minutes but is more typically 2 to 3 minutes in length.

Total eclipses all begin and end with a series of partial phases lasting an hour or more. But this is where the resemblance to partial and annular eclipses ends — the total phase is the most spectacular astronomical event visible to the naked eye. During totality, the Sun's outer atmosphere — the solar corona — appears as a gossamer halo surrounding the Moon, and bright stars and planets are visible.

The eclipse takes on a unique character about five minutes before the total phase commences. The Sun has a foreboding quality and casts abnormally sharp shadows. The approaching lunar umbra darkens the western sky and the air temperature is noticeably cooler. A minute before totality, pale shadow bands[2] ripple across the ground.

The ambient light grows feeble even though the crescent Sun is still too bright to see. In the final seconds, the Sun's corona emerges from the glare as the solar crescent shrinks to a brilliant jewel. This celestial diamond ring lingers for a moment before the sunlight is extinguished and totality begins.

The Sun's glorious corona is now displayed to full advantage in the darkened sky. Standing within the Moon's umbra affords the rare and unprecedented opportunity to gaze directly at the glowing million-degree plasma surrounding our star. Twisted, tortured, and constrained by the Sun's enormous magnetic fields, the solar corona is revealed to the naked eye only during the brief seconds when the Moon completely blocks the brilliant disk of the Sun. An eerie twilight bathes the landscape and the colors of dusk surround the horizon.

Minutes race by like seconds. Suddenly, a sparkling bead of sunlight reappears along one edge of the Moon and quickly grows to blindingly bright proportions. Daylight returns as the corona fades and the totality ends. Another hour of partial phases remains before the eclipse is over.

While filters are required for viewing the partial phases, they must be removed for totality. The total phase is the only time it is completely safe to look directly at the Sun without protection. In fact, the total phase is not even visible through solar filters because the Sun's corona is a million times fainter than the photosphere[3].

Figure 1–3: Solar Eclipse Contacts

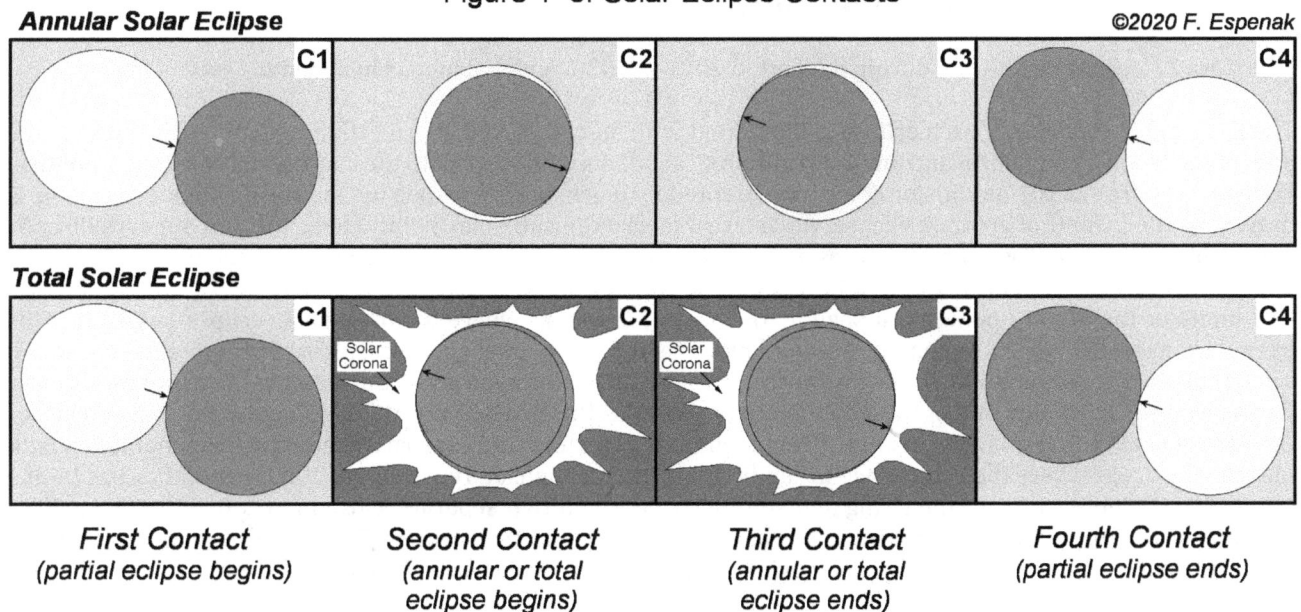

Figure 1–3 illustrates the four contacts for annular and total solar eclipses. The arrows indicate the contact point of the Moon with the Sun's limb in each diagram.

Solar Eclipse Contacts

During the course of a solar eclipse, the instants when the edge of the Moon's disk becomes tangent to the Sun's disk are known as eclipse contacts. They mark various stages or phases of a solar eclipse (Figure 1–3).

Partial solar eclipses have two primary contacts.

First Contact (C1) — Partial Eclipse Begins (Instant of first exterior tangency of the Moon with the Sun)
Fourth Contact (C4) — Partial Eclipse Ends (Instant of last exterior tangency of the Moon with the Sun)

---

[2] Shadow bands are caused by the thin solar crescent illuminating Earth's atmosphere before and after totality.
[3] The photosphere is the visible surface of the Sun's disk.

Central solar eclipses (total, annular or hybrid) have four primary contacts. Contacts C2 and C3 mark the instants when the Moon's disk is first and last internally tangent to the Sun. These are the times when the annular or total phase of the eclipse begins and ends, respectively.

First Contact (C1) — Partial Eclipse Begins (Instant of first exterior tangency of the Moon with the Sun)
Second Contact (C2) — Total (or Annular) Eclipse Begins (Instant of first interior tangency of Moon with Sun)
Third Contact (C3) — Total (or Annular) Eclipse Ends (Instant of last interior tangency of Moon with Sun)
Fourth Contact (C4) — Partial Eclipse Ends (Instant of last exterior tangency of the Moon with the Sun)

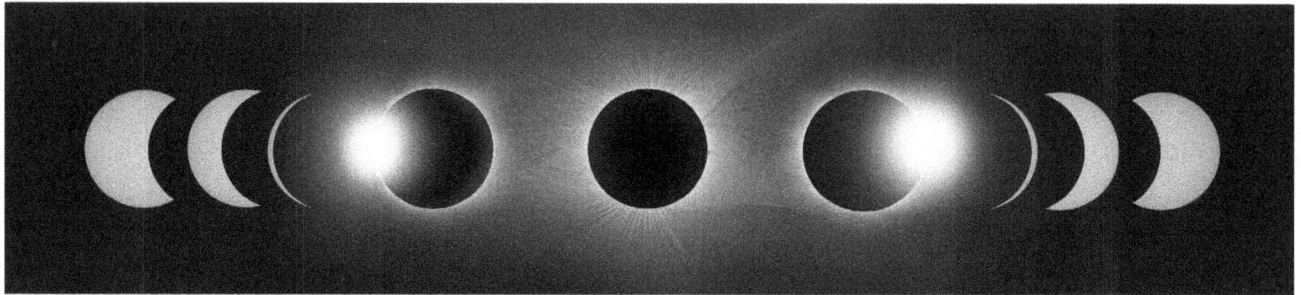

Photo 1–6 Sequence of partial phases, Diamond Rings, and Totality. Total Solar Eclipse of 2017 Aug 21. ©2017 F. Espenak

Explanation of Global Solar Eclipse Maps

There are 22 eclipses of the Sun during the period 2021 to 2030. A global map is included for each eclipse.

The geographic visibility of each eclipse is illustrated with an orthographic projection map of Earth showing the path of the Moon's penumbral and umbral/antumbral shadows with respect to the continental coastlines, political boundaries (circa 2016), and major cities (represented by black dots). North is up and the daylight terminator is drawn for the instant of greatest eclipse. An asterisk marks the sub-solar point where the Sun appears directly overhead at that time. The salient features of the eclipse maps are identified in the key diagram on page 11.

The limits of the Moon's penumbral shadow delineate the region where a partial solar eclipse is visible. This irregular or saddle shaped region often covers more than half the daylight hemisphere of Earth and consists of several distinct zones or limits. At the northern and/or southern boundaries lie the limits of the penumbra's path. Partial eclipses have only one of these limits, as do central eclipses when the Moon's shadow axis falls no closer than about 0.45 radii from Earth's center. Great loops at the western and eastern extremes of the penumbra's path identify the areas where the eclipse begins/ends at sunrise and sunset, respectively. If the penumbra has both a northern and southern limit, the rising and setting curves form two separate, closed loops (e.g., 2024 Apr 08). Otherwise, the curves are connected in a distorted figure eight (e.g., 2021 Dec 04). Bisecting the *eclipse begins/ends at sunrise and sunset* loops is the curve of maximum eclipse at sunrise (western loop) and sunset (eastern loop).

The eclipse magnitude is defined as the fraction of the Sun's diameter occulted by the Moon. A curve of constant eclipse magnitude delineates the points where the local magnitude at maximum eclipse is equal to a constant value. The maps include *curves of constant eclipse magnitude* for values of 0.2, 0.4, 0.6, and 0.8. These curves run exclusively between the curves of maximum eclipse at sunrise and sunset. They are approximately parallel to the northern/southern penumbral limits and the umbral/antumbral paths of central eclipses. The northern and southern limits of the penumbra may be thought of as curves of eclipse magnitude of 0.0. For total eclipses, the northern and southern limits of the umbra are curves of eclipse magnitude of 1.0.

Greatest eclipse is the instant when the axis of the Moon's shadow cone passes closest to Earth's center. The point on Earth's surface intersected by the axis of the Moon's shadow cone at greatest eclipse is marked by an asterisk. For partial eclipses, the shadow axis misses Earth entirely, so the point of greatest eclipse lies on the day/night terminator and the Sun appears on the horizon.

Parameters for the eclipse appears to the right of each map. They include the time of greatest eclipse (Universal Time[4] or UT1), the eclipse magnitude and obscuration, gamma[5], Saros series, and node (ascending or descending).

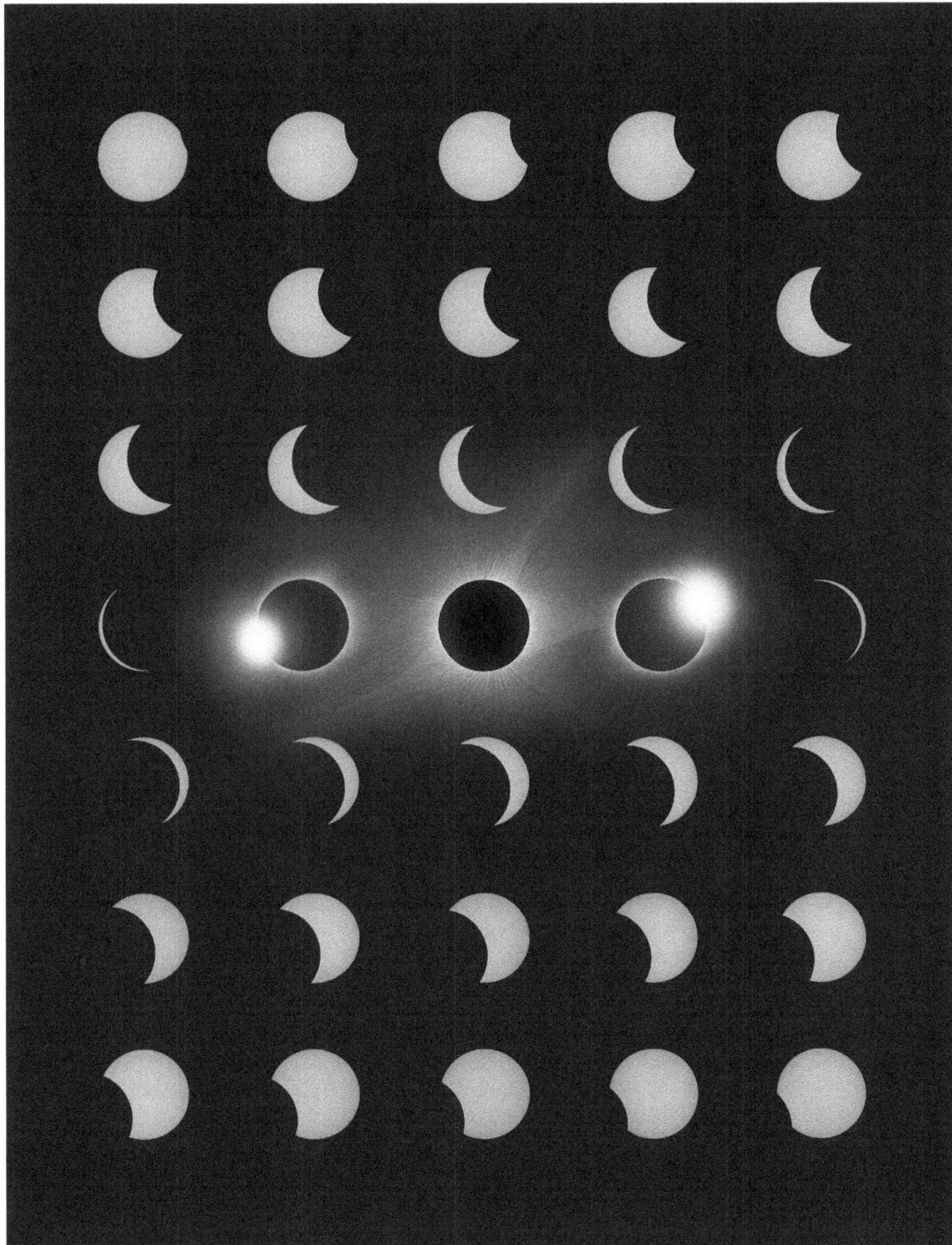

Photo 1–7 Time Sequence of the Total Solar Eclipse of 2017 Aug 21. ©2017 F. Espenak

---

[4] Universal Time (UT1) is the modern-day replacement for Greenwich Mean Time.
[5] Gamma is the distance of the Moon's shadow cone axis from Earth's center at greatest eclipse (in Earth equatorial radii).

Global Solar Eclipse Maps

Key to Global Solar Eclipse Maps

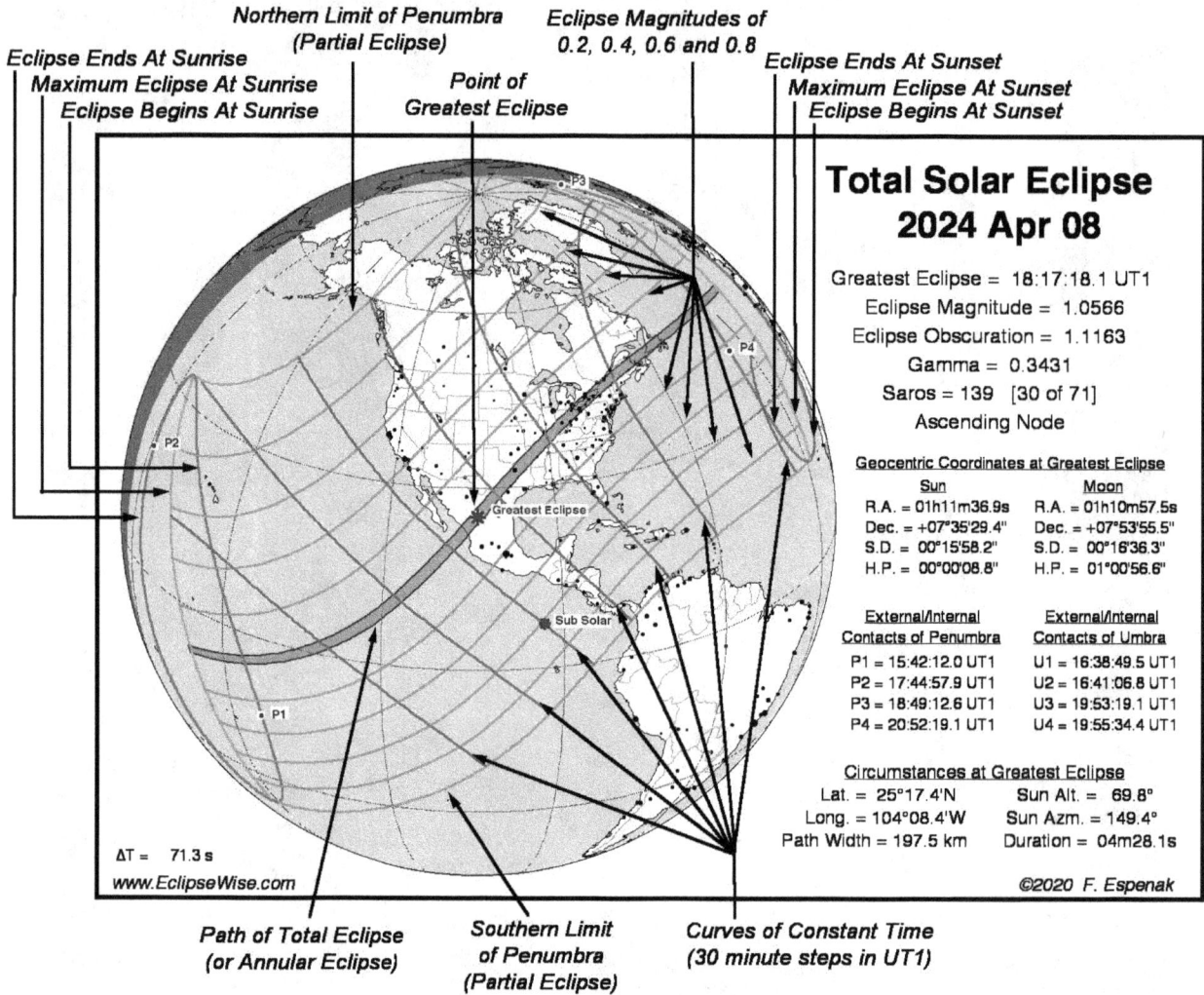

Eclipse Ends At Sunrise
Maximum Eclipse At Sunrise
Eclipse Begins At Sunrise

Northern Limit of Penumbra
(Partial Eclipse)

Point of
Greatest Eclipse

Eclipse Magnitudes of
0.2, 0.4, 0.6 and 0.8

Eclipse Ends At Sunset
Maximum Eclipse At Sunset
Eclipse Begins At Sunset

**Total Solar Eclipse
2024 Apr 08**

Greatest Eclipse = 18:17:18.1 UT1
Eclipse Magnitude = 1.0566
Eclipse Obscuration = 1.1163
Gamma = 0.3431
Saros = 139 [30 of 71]
Ascending Node

Geocentric Coordinates at Greatest Eclipse

	Sun	Moon
R.A. =	01h11m36.9s	01h10m57.5s
Dec. =	+07°35'29.4"	+07°53'55.5"
S.D. =	00°15'58.2"	00°16'36.3"
H.P. =	00°00'08.8"	01°00'56.6"

External/Internal Contacts of Penumbra
P1 = 15:42:12.0 UT1
P2 = 17:44:57.9 UT1
P3 = 18:49:12.6 UT1
P4 = 20:52:19.1 UT1

External/Internal Contacts of Umbra
U1 = 16:38:49.5 UT1
U2 = 16:41:06.8 UT1
U3 = 19:53:19.1 UT1
U4 = 19:55:34.4 UT1

Circumstances at Greatest Eclipse
Lat. = 25°17.4'N   Sun Alt. = 69.8°
Long. = 104°08.4'W   Sun Azm. = 149.4°
Path Width = 197.5 km Duration = 04m28.1s

ΔT = 71.3 s
www.EclipseWise.com

©2020 F. Espenak

Path of Total Eclipse
(or Annular Eclipse)

Southern Limit
of Penumbra
(Partial Eclipse)

Curves of Constant Time
(30 minute steps in UT1)

Explanation of Terms Used in Global Solar Eclipse Maps

Greatest Eclipse – The instant when the distance between the axis of the Moon's shadow cone and the center of Earth reaches a minimum (in Universal Time[6] or UT1).

Eclipse Magnitude – The fraction of the Sun's diameter occulted by the Moon at the instant of greatest eclipse (for total and annular eclipses this value is the ratio of diameters of the Moon and the Sun).

Eclipse Obscuration – The fraction of the Sun's area occulted by the Moon at the instant of greatest eclipse.

Gamma – The minimum distance from the lunar shadow axis to the center of Earth (units of Earth equatorial radii).

Saros Series – The Saros series that the eclipse belongs to. The numbers in "[ ]" are the eclipse's sequential position and the number of eclipses in the Saros series.

Node – The orbital node near which the eclipse takes place (Ascending Node or Descending Node).

Geocentric Coordinates of the Sun and the Moon at Greatest Eclipse

R.A. – Right Ascension   S.D. – Semi-Diameter (i.e. - radius)
Dec. – Declination   H.P. – Horizontal Parallax

External/Internal Contacts of Penumbra and Umbra – Instants when each shadow enters or exits the surface of Earth (Penumbral contacts shown on map as: P1, P2, P3, and P4; umbral contacts are located at the ends of the central path). All times are in Universal Time (UT1)

Circumstances of Greatest Eclipse – Geographic location (Lat., Long.) of the shadow axis, the altitude and azimuth of the Sun, width of the central path, and the central duration of totality or annularity.

---

[6] Universal Time or UT1 is the modern replacement for Greenwich Mean Time

Annular Solar Eclipse
2021 Jun 10

Greatest Eclipse = 10:41:56.6 UT1

Eclipse Magnitude = 0.9435

Eclipse Obscuration = 0.8902

Gamma = 0.9152

Saros = 147 [23 of 80]

Ascending Node

Geocentric Coordinates at Greatest Eclipse

Sun	Moon
R.A. = 05h15m31.4s	R.A. = 05h14m53.6s
Dec. = +23°02'37.1"	Dec. = +23°51'21.6"
S.D. = 00°15'45.2"	S.D. = 00°14'46.8"
H.P. = 00°00'08.7"	H.P. = 00°54'14.5"

External/Internal Contacts of Penumbra	External/Internal Contacts of Umbra
P1 = 08:12:20.5 UT1	U1 = 09:49:48.1 UT1
P4 = 13:11:21.8 UT1	U2 = 10:00:41.3 UT1
	U3 = 11:23:00.3 UT1
	U4 = 11:33:51.2 UT1

Circumstances at Greatest Eclipse

Lat. = 80°48.9'N	Sun Alt. = 23.3°
Long. = 066°46.1'W	Sun Azm. = 89.9°
Path Width = 526.8 km	Duration = 03m51.2s

ΔT = 70.1 s

www.EclipseWise.com

©2020 F. Espenak

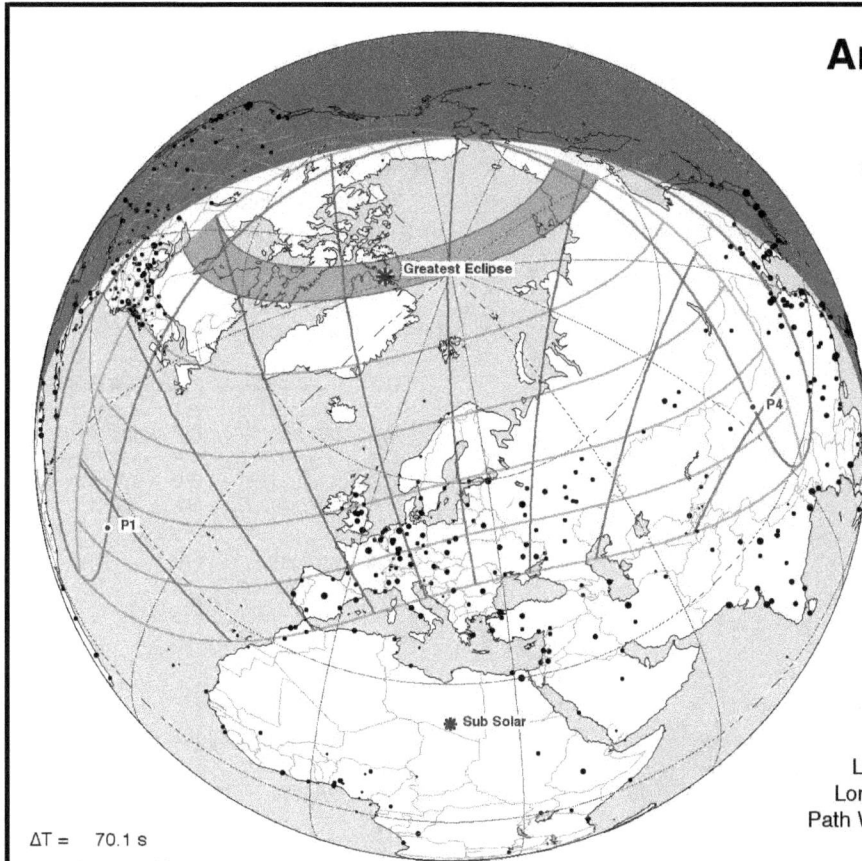

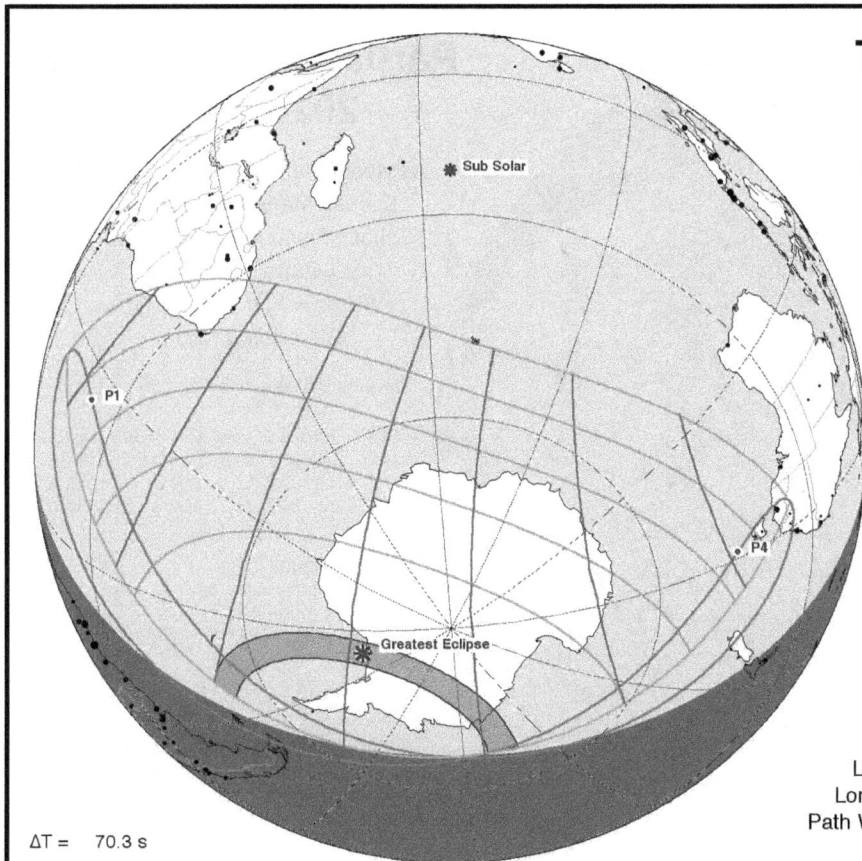

Total Solar Eclipse
2021 Dec 04

Greatest Eclipse = 07:33:27.6 UT1

Eclipse Magnitude = 1.0367

Eclipse Obscuration = 1.0748

Gamma = -0.9526

Saros = 152 [13 of 70]

Decending Node

Geocentric Coordinates at Greatest Eclipse

Sun	Moon
R.A. = 16h43m32.4s	R.A. = 16h42m35.0s
Dec. = -22°16'29.4"	Dec. = -23°13'22.3"
S.D. = 00°16'13.6"	S.D. = 00°16'44.7"
H.P. = 00°00'08.9"	H.P. = 01°01'27.3"

External/Internal Contacts of Penumbra	External/Internal Contacts of Umbra
P1 = 05:29:16.3 UT1	U1 = 07:00:06.4 UT1
P4 = 09:37:29.1 UT1	U2 = 07:05:54.5 UT1
	U3 = 08:00:45.5 UT1
	U4 = 08:06:34.1 UT1

Circumstances at Greatest Eclipse

Lat. = 76°46.6'S	Sun Alt. = 17.2°
Long. = 046°13.8'W	Sun Azm. = 114.8°
Path Width = 418.8 km	Duration = 01m54.4s

ΔT = 70.3 s

www.EclipseWise.com

©2020 F. Espenak

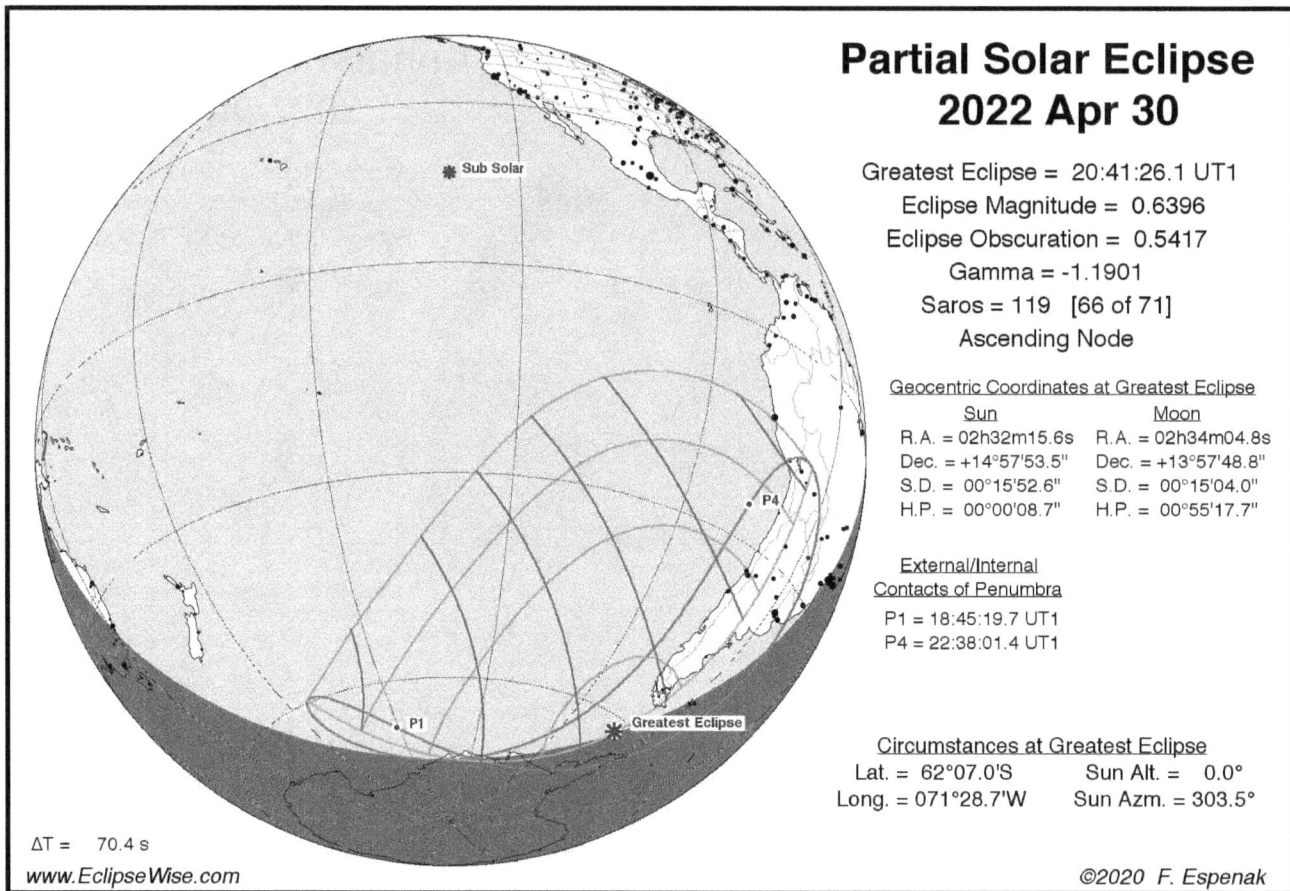

Partial Solar Eclipse
2022 Apr 30

Greatest Eclipse = 20:41:26.1 UT1

Eclipse Magnitude = 0.6396

Eclipse Obscuration = 0.5417

Gamma = -1.1901

Saros = 119 [66 of 71]

Ascending Node

Geocentric Coordinates at Greatest Eclipse

Sun	Moon
R.A. = 02h32m15.6s	R.A. = 02h34m04.8s
Dec. = +14°57'53.5"	Dec. = +13°57'48.8"
S.D. = 00°15'52.6"	S.D. = 00°15'04.0"
H.P. = 00°00'08.7"	H.P. = 00°55'17.7"

External/Internal
Contacts of Penumbra

P1 = 18:45:19.7 UT1
P4 = 22:38:01.4 UT1

Circumstances at Greatest Eclipse

Lat. = 62°07.0'S	Sun Alt. = 0.0°
Long. = 071°28.7'W	Sun Azm. = 303.5°

ΔT = 70.4 s

www.EclipseWise.com

©2020 F. Espenak

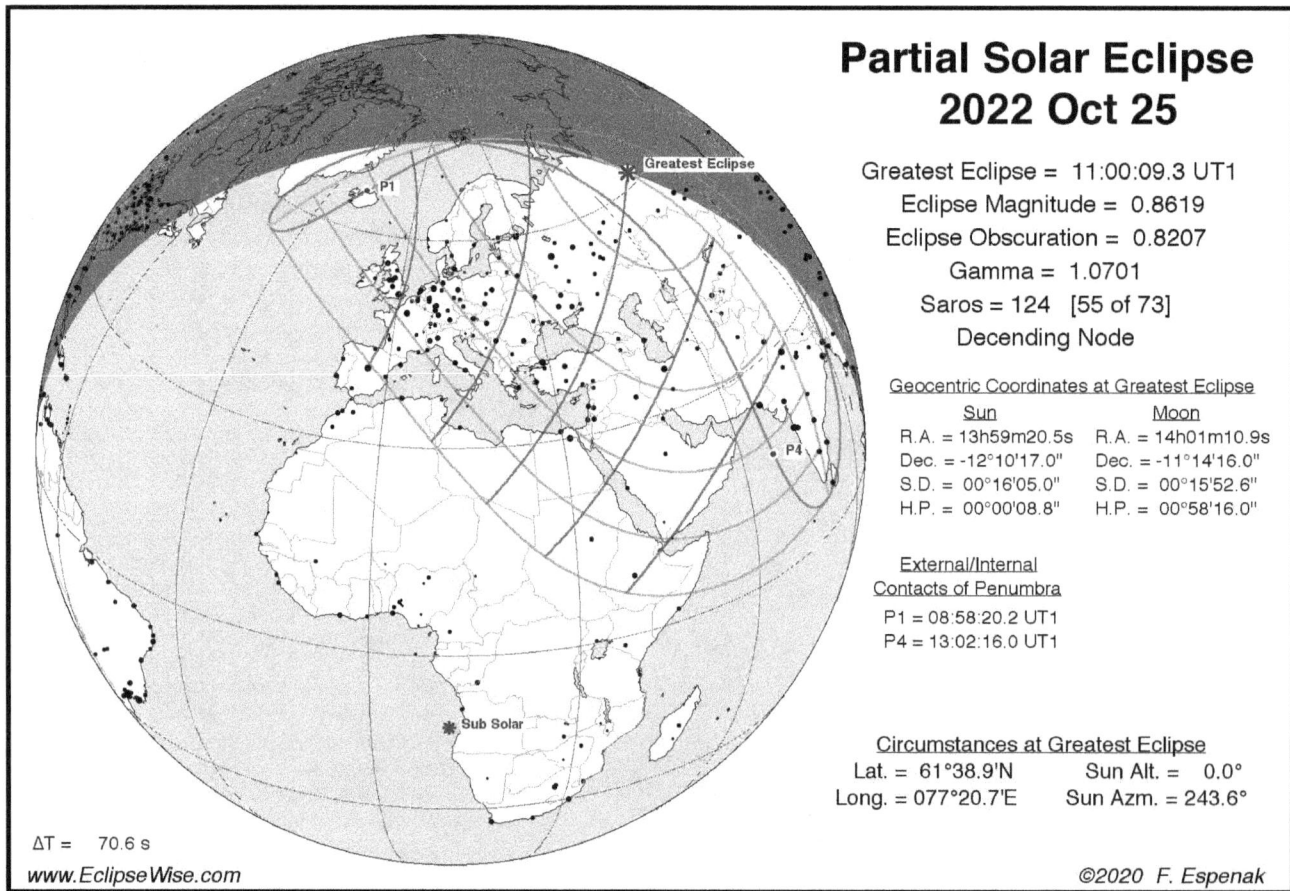

Partial Solar Eclipse
2022 Oct 25

Greatest Eclipse = 11:00:09.3 UT1

Eclipse Magnitude = 0.8619

Eclipse Obscuration = 0.8207

Gamma = 1.0701

Saros = 124 [55 of 73]

Decending Node

Geocentric Coordinates at Greatest Eclipse

Sun	Moon
R.A. = 13h59m20.5s	R.A. = 14h01m10.9s
Dec. = -12°10'17.0"	Dec. = -11°14'16.0"
S.D. = 00°16'05.0"	S.D. = 00°15'52.6"
H.P. = 00°00'08.8"	H.P. = 00°58'16.0"

External/Internal
Contacts of Penumbra

P1 = 08:58:20.2 UT1
P4 = 13:02:16.0 UT1

Circumstances at Greatest Eclipse

Lat. = 61°38.9'N	Sun Alt. = 0.0°
Long. = 077°20.7'E	Sun Azm. = 243.6°

ΔT = 70.6 s

www.EclipseWise.com

©2020 F. Espenak

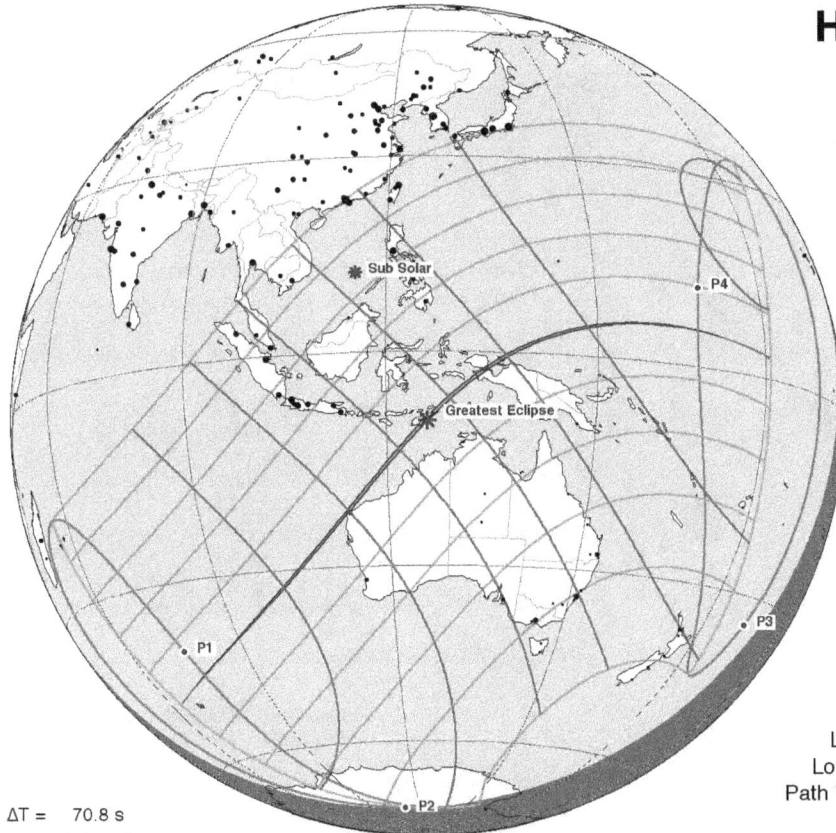

Hybrid Solar Eclipse
2023 Apr 20

Greatest Eclipse = 04:16:45.1 UT1
Eclipse Magnitude = 1.0132
Eclipse Obscuration = 1.0266
Gamma = -0.3952
Saros = 129   [52 of 80]
Ascending Node

Geocentric Coordinates at Greatest Eclipse

Sun	Moon
R.A. = 01h51m01.7s	R.A. = 01h51m43.2s
Dec. = +11°24'54.1"	Dec. = +11°04'16.7"
S.D. = 00°15'55.4"	S.D. = 00°15'53.6"
H.P. = 00°00'08.8"	H.P. = 00°58'19.9"

External/Internal Contacts of Penumbra	External/Internal Contacts of Umbra
P1 = 01:34:23.5 UT1	U1 = 02:37:04.1 UT1
P2 = 03:53:21.4 UT1	U2 = 02:37:10.9 UT1
P3 = 04:40:37.0 UT1	U3 = 05:56:30.5 UT1
P4 = 06:59:21.1 UT1	U4 = 05:56:42.6 UT1

Circumstances at Greatest Eclipse

Lat. = 09°35.7'S	Sun Alt. = 66.7°
Long. = 125°46.8'E	Sun Azm. = 334.0°
Path Width = 49.0 km	Duration = 01m16.1s

ΔT = 70.8 s

www.EclipseWise.com

©2020 F. Espenak

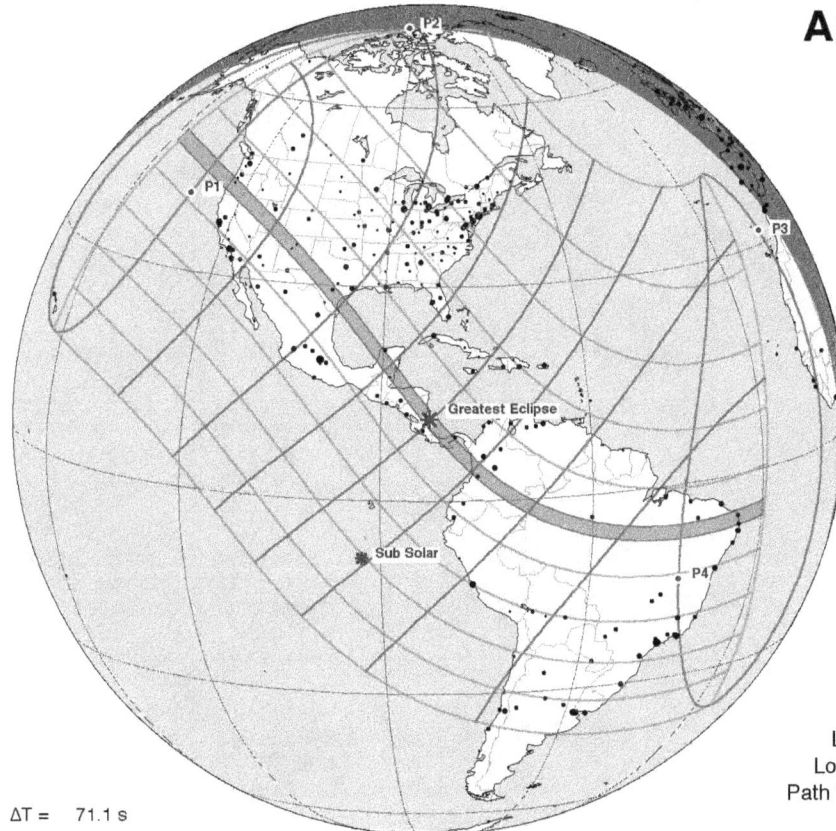

Annular Solar Eclipse
2023 Oct 14

Greatest Eclipse = 17:59:29.5 UT1
Eclipse Magnitude = 0.9520
Eclipse Obscuration = 0.9064
Gamma = 0.3753
Saros = 134   [44 of 71]
Decending Node

Geocentric Coordinates at Greatest Eclipse

Sun	Moon
R.A. = 13h18m05.4s	R.A. = 13h18m44.3s
Dec. = -08°14'36.7"	Dec. = -07°56'18.9"
S.D. = 00°16'02.0"	S.D. = 00°15'02.9"
H.P. = 00°00'08.8"	H.P. = 00°55'13.8"

External/Internal Contacts of Penumbra	External/Internal Contacts of Umbra
P1 = 15:03:47.1 UT1	U1 = 16:10:08.0 UT1
P2 = 17:34:38.8 UT1	U2 = 16:14:41.4 UT1
P3 = 18:24:54.0 UT1	U3 = 19:44:33.9 UT1
P4 = 20:55:15.7 UT1	U4 = 19:49:02.0 UT1

Circumstances at Greatest Eclipse

Lat. = 11°22.1'N	Sun Alt. = 67.9°
Long. = 083°06.2'W	Sun Azm. = 208.0°
Path Width = 187.4 km	Duration = 05m17.2s

ΔT = 71.1 s

www.EclipseWise.com

©2020 F. Espenak

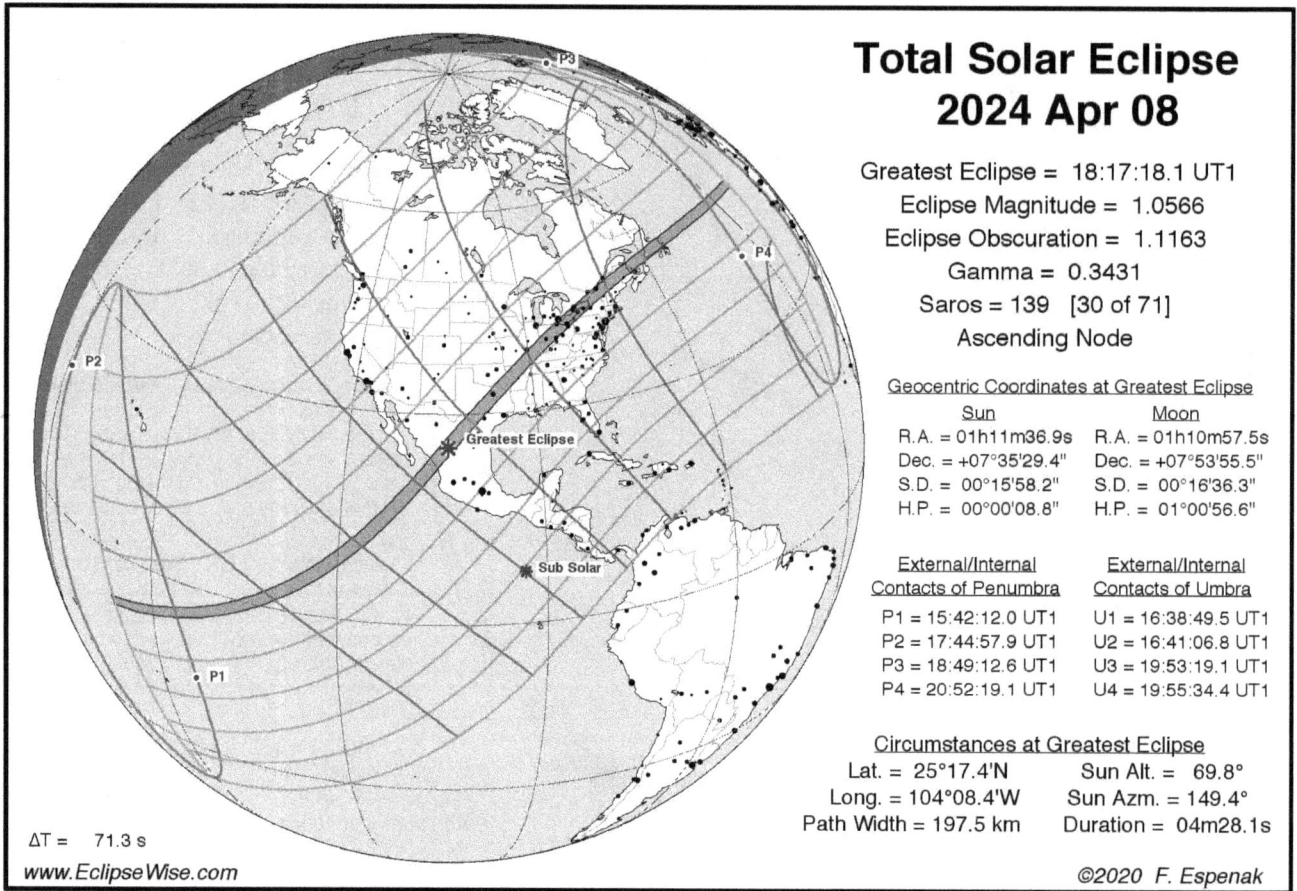

Total Solar Eclipse
2024 Apr 08

Greatest Eclipse = 18:17:18.1 UT1
Eclipse Magnitude = 1.0566
Eclipse Obscuration = 1.1163
Gamma = 0.3431
Saros = 139 [30 of 71]
Ascending Node

Geocentric Coordinates at Greatest Eclipse

	Sun	Moon
R.A. =	01h11m36.9s	R.A. = 01h10m57.5s
Dec. =	+07°35'29.4"	Dec. = +07°53'55.5"
S.D. =	00°15'58.2"	S.D. = 00°16'36.3"
H.P. =	00°00'08.8"	H.P. = 01°00'56.6"

External/Internal Contacts of Penumbra	External/Internal Contacts of Umbra
P1 = 15:42:12.0 UT1	U1 = 16:38:49.5 UT1
P2 = 17:44:57.9 UT1	U2 = 16:41:06.8 UT1
P3 = 18:49:12.6 UT1	U3 = 19:53:19.1 UT1
P4 = 20:52:19.1 UT1	U4 = 19:55:34.4 UT1

Circumstances at Greatest Eclipse

Lat. = 25°17.4'N	Sun Alt. = 69.8°
Long. = 104°08.4'W	Sun Azm. = 149.4°
Path Width = 197.5 km	Duration = 04m28.1s

ΔT = 71.3 s

www.EclipseWise.com

©2020 F. Espenak

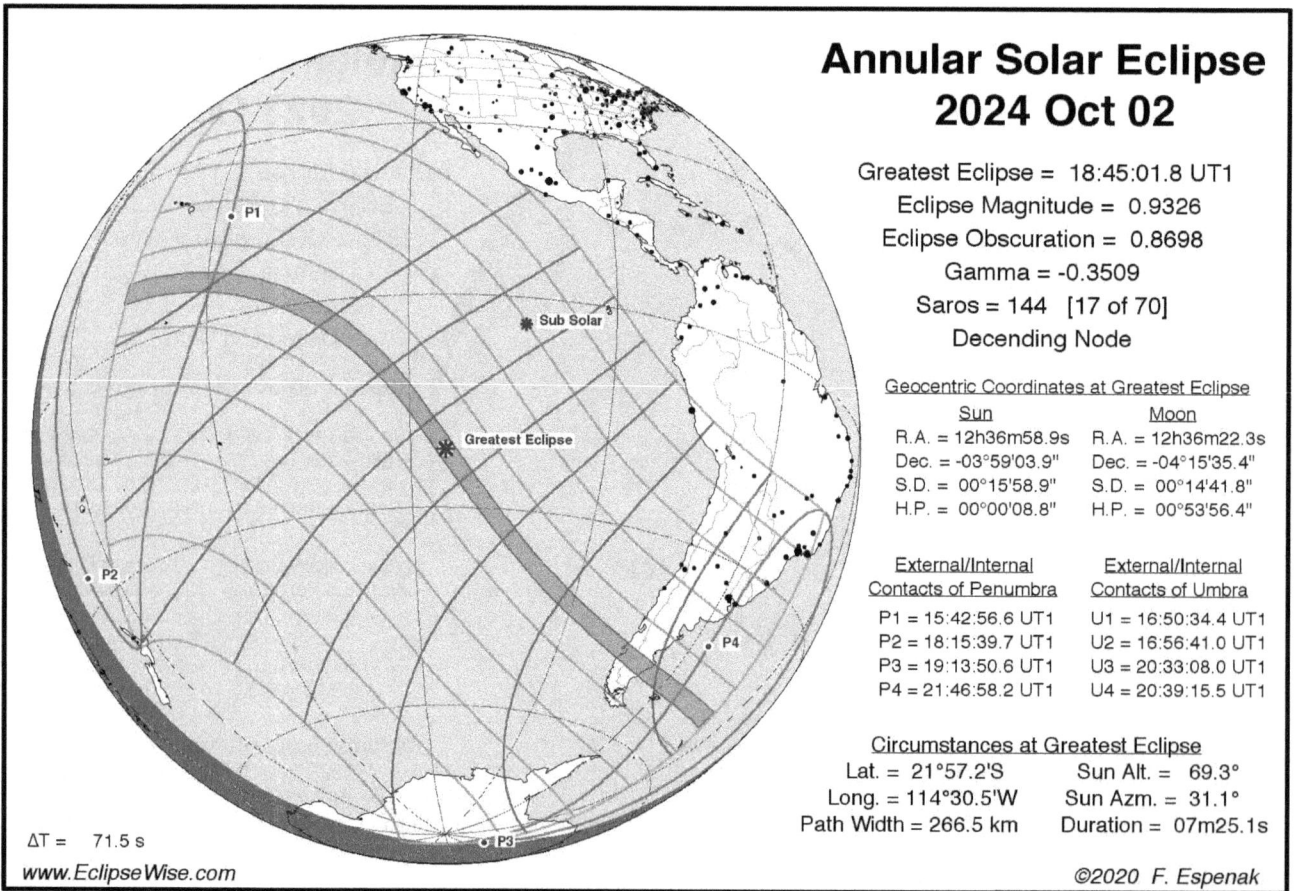

Annular Solar Eclipse
2024 Oct 02

Greatest Eclipse = 18:45:01.8 UT1
Eclipse Magnitude = 0.9326
Eclipse Obscuration = 0.8698
Gamma = -0.3509
Saros = 144 [17 of 70]
Decending Node

Geocentric Coordinates at Greatest Eclipse

	Sun	Moon
R.A. =	12h36m58.9s	R.A. = 12h36m22.3s
Dec. =	-03°59'03.9"	Dec. = -04°15'35.4"
S.D. =	00°15'58.9"	S.D. = 00°14'41.8"
H.P. =	00°00'08.8"	H.P. = 00°53'56.4"

External/Internal Contacts of Penumbra	External/Internal Contacts of Umbra
P1 = 15:42:56.6 UT1	U1 = 16:50:34.4 UT1
P2 = 18:15:39.7 UT1	U2 = 16:56:41.0 UT1
P3 = 19:13:50.6 UT1	U3 = 20:33:08.0 UT1
P4 = 21:46:58.2 UT1	U4 = 20:39:15.5 UT1

Circumstances at Greatest Eclipse

Lat. = 21°57.2'S	Sun Alt. = 69.3°
Long. = 114°30.5'W	Sun Azm. = 31.1°
Path Width = 266.5 km	Duration = 07m25.1s

ΔT = 71.5 s

www.EclipseWise.com

©2020 F. Espenak

Partial Solar Eclipse
2025 Mar 29

Greatest Eclipse = 10:47:24.4 UT1
Eclipse Magnitude = 0.9376
Eclipse Obscuration = 0.9306
Gamma = 1.0405
Saros = 149 [21 of 71]
Ascending Node

Geocentric Coordinates at Greatest Eclipse

Sun	Moon
R.A. = 00h33m03.1s	R.A. = 00h31m00.8s
Dec. = +03°33'55.0"	Dec. = +04°29'34.1"
S.D. = 00°16'01.1"	S.D. = 00°16'39.4"
H.P. = 00°00'08.8"	H.P. = 01°01'07.8"

External/Internal
Contacts of Penumbra
P1 = 08:50:40.8 UT1
P4 = 12:43:42.3 UT1

Circumstances at Greatest Eclipse

Lat. = 61°06.0'N	Sun Alt. = 0.0°
Long. = 077°05.1'W	Sun Azm. = 82.6°

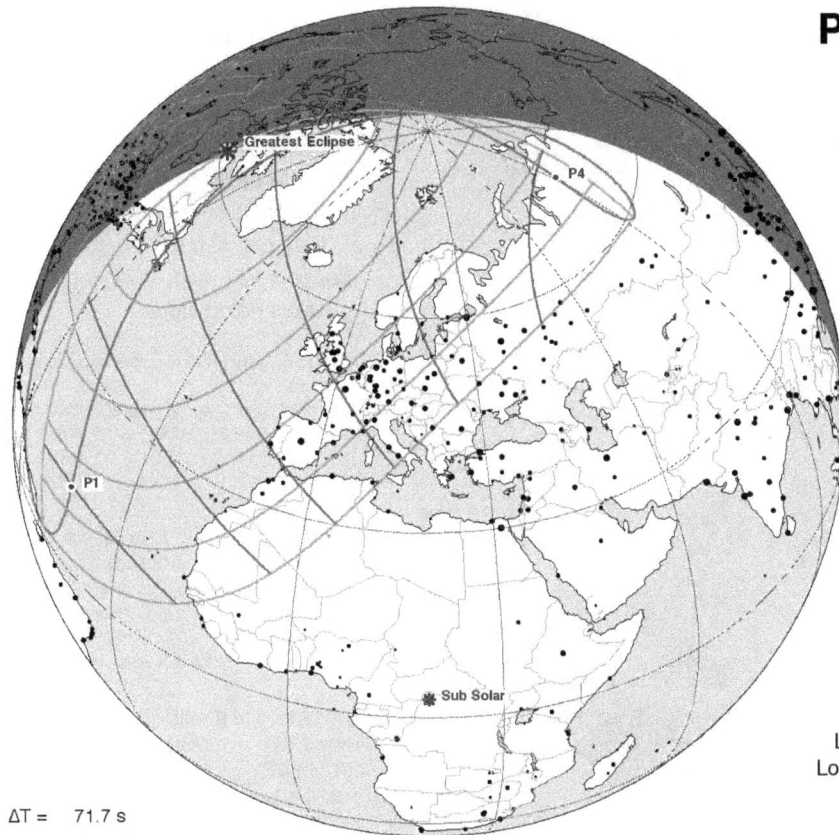

ΔT = 71.7 s

www.EclipseWise.com

©2020 F. Espenak

Partial Solar Eclipse
2025 Sep 21

Greatest Eclipse = 19:41:52.3 UT1
Eclipse Magnitude = 0.8550
Eclipse Obscuration = 0.7969
Gamma = -1.0651
Saros = 154 [7 of 71]
Decending Node

Geocentric Coordinates at Greatest Eclipse

Sun	Moon
R.A. = 11h56m36.9s	R.A. = 11h54m42.8s
Dec. = +00°22'00.7"	Dec. = -00°29'14.7"
S.D. = 00°15'55.9"	S.D. = 00°15'02.8"
H.P. = 00°00'08.8"	H.P. = 00°55'13.2"

External/Internal
Contacts of Penumbra
P1 = 17:29:39.4 UT1
P4 = 21:53:43.2 UT1

Circumstances at Greatest Eclipse

Lat. = 60°54.1'S	Sun Alt. = 0.0°
Long. = 153°29.9'E	Sun Azm. = 89.2°

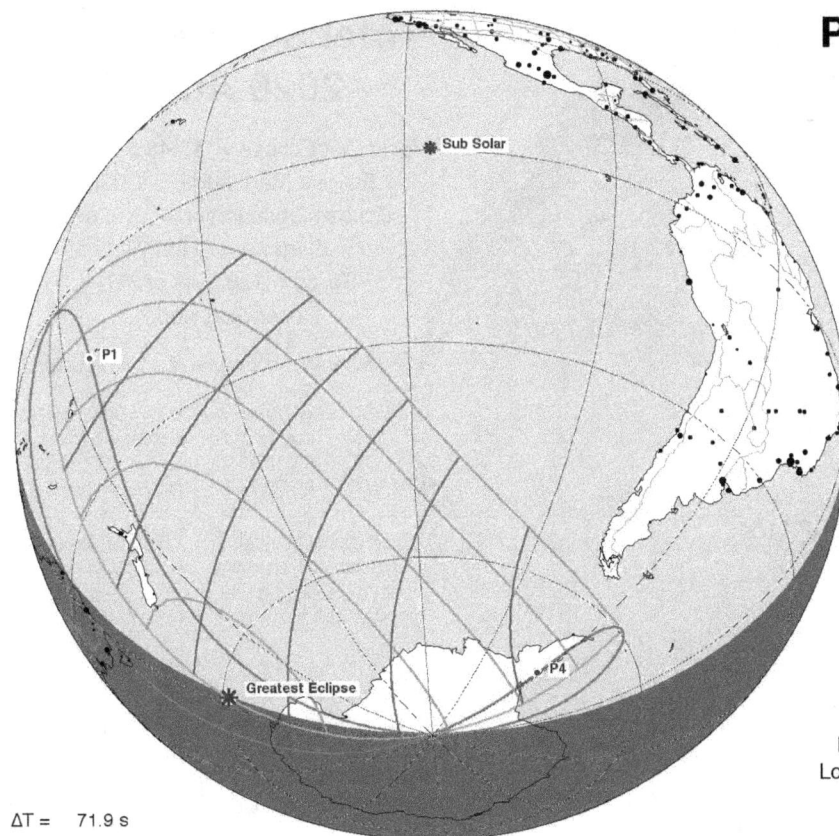

ΔT = 71.9 s

www.EclipseWise.com

©2020 F. Espenak

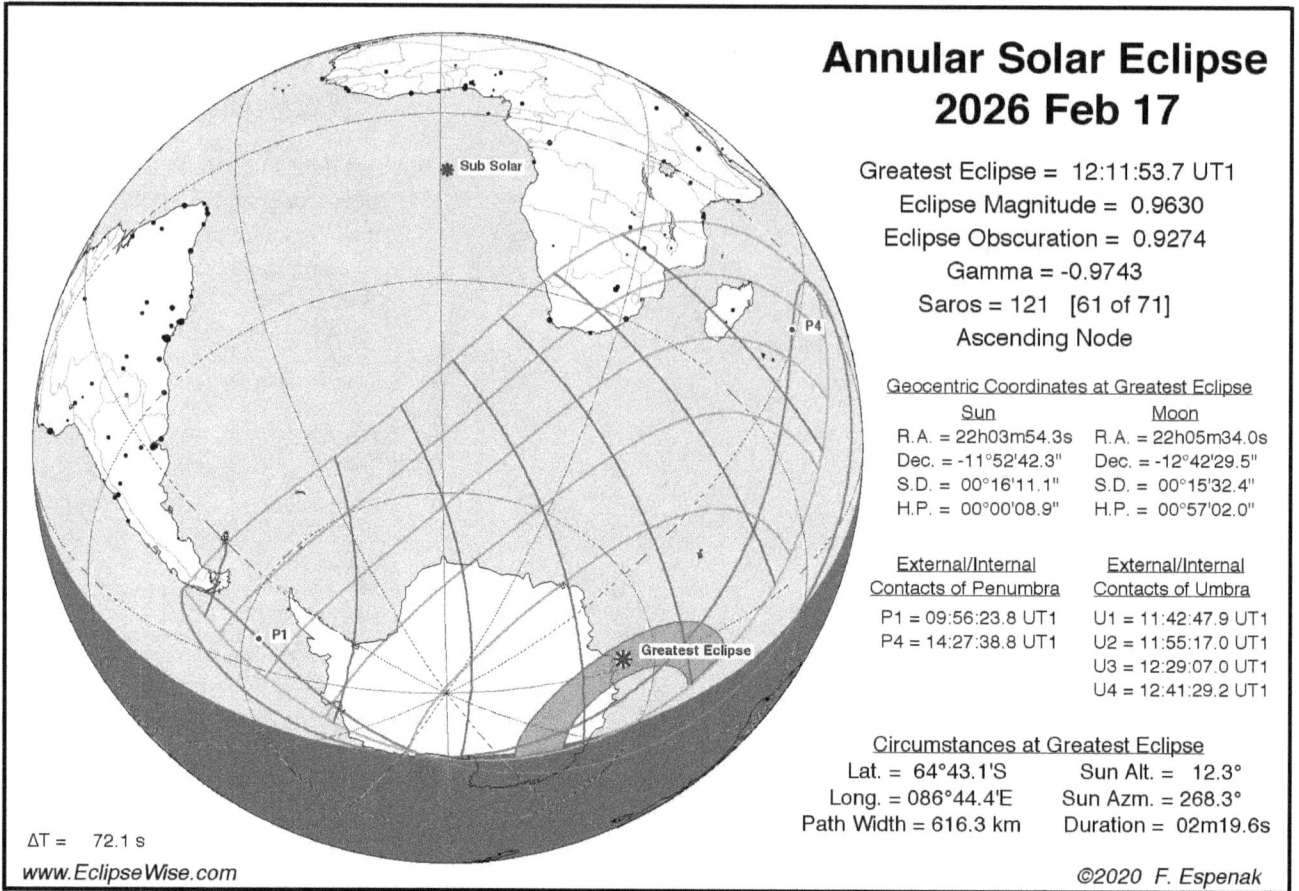

Annular Solar Eclipse
2026 Feb 17

Greatest Eclipse = 12:11:53.7 UT1
Eclipse Magnitude = 0.9630
Eclipse Obscuration = 0.9274
Gamma = -0.9743
Saros = 121 [61 of 71]
Ascending Node

Geocentric Coordinates at Greatest Eclipse

Sun	Moon
R.A. = 22h03m54.3s	R.A. = 22h05m34.0s
Dec. = -11°52'42.3"	Dec. = -12°42'29.5"
S.D. = 00°16'11.1"	S.D. = 00°15'32.4"
H.P. = 00°00'08.9"	H.P. = 00°57'02.0"

External/Internal Contacts of Penumbra	External/Internal Contacts of Umbra
P1 = 09:56:23.8 UT1	U1 = 11:42:47.9 UT1
P4 = 14:27:38.8 UT1	U2 = 11:55:17.0 UT1
	U3 = 12:29:07.0 UT1
	U4 = 12:41:29.2 UT1

Circumstances at Greatest Eclipse

Lat. = 64°43.1'S	Sun Alt. = 12.3°
Long. = 086°44.4'E	Sun Azm. = 268.3°
Path Width = 616.3 km	Duration = 02m19.6s

ΔT = 72.1 s

www.EclipseWise.com

©2020 F. Espenak

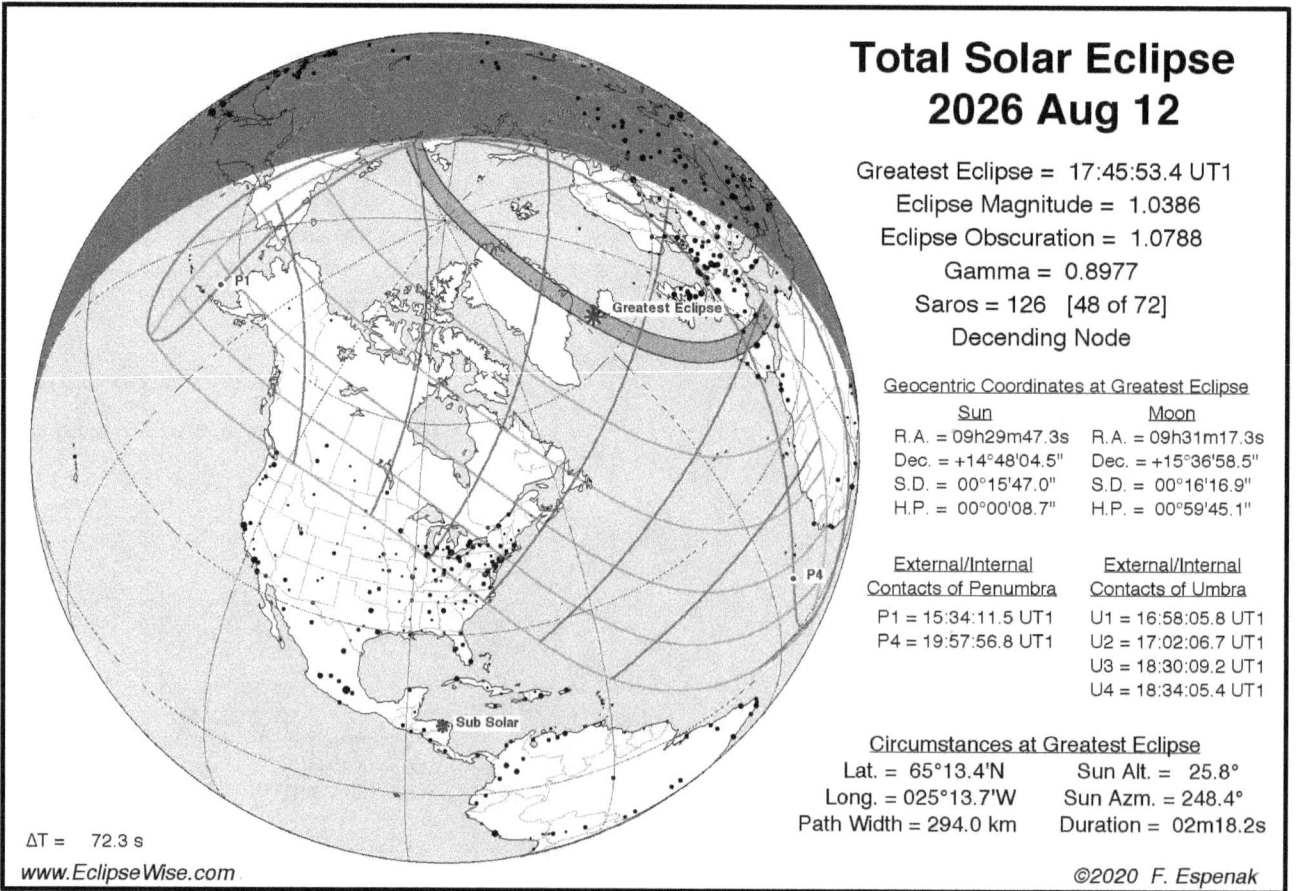

Total Solar Eclipse
2026 Aug 12

Greatest Eclipse = 17:45:53.4 UT1
Eclipse Magnitude = 1.0386
Eclipse Obscuration = 1.0788
Gamma = 0.8977
Saros = 126 [48 of 72]
Decending Node

Geocentric Coordinates at Greatest Eclipse

Sun	Moon
R.A. = 09h29m47.3s	R.A. = 09h31m17.3s
Dec. = +14°48'04.5"	Dec. = +15°36'58.5"
S.D. = 00°15'47.0"	S.D. = 00°16'16.9"
H.P. = 00°00'08.7"	H.P. = 00°59'45.1"

External/Internal Contacts of Penumbra	External/Internal Contacts of Umbra
P1 = 15:34:11.5 UT1	U1 = 16:58:05.8 UT1
P4 = 19:57:56.8 UT1	U2 = 17:02:06.7 UT1
	U3 = 18:30:09.2 UT1
	U4 = 18:34:05.4 UT1

Circumstances at Greatest Eclipse

Lat. = 65°13.4'N	Sun Alt. = 25.8°
Long. = 025°13.7'W	Sun Azm. = 248.4°
Path Width = 294.0 km	Duration = 02m18.2s

ΔT = 72.3 s

www.EclipseWise.com

©2020 F. Espenak

Annular Solar Eclipse
2027 Feb 06

Greatest Eclipse = 15:59:35.2 UT1

Eclipse Magnitude = 0.9281

Eclipse Obscuration = 0.8614

Gamma = -0.2952

Saros = 131   [51 of 70]

Ascending Node

Geocentric Coordinates at Greatest Eclipse

Sun	Moon
R.A. = 21h20m17.6s	R.A. = 21h20m44.2s
Dec. = -15°32'54.5"	Dec. = -15°47'36.0"
S.D. = 00°16'13.1"	S.D. = 00°14'50.2"
H.P. = 00°00'08.9"	H.P. = 00°54'27.0"

External/Internal Contacts of Penumbra	External/Internal Contacts of Umbra
P1 = 12:57:34.5 UT1	U1 = 14:03:53.0 UT1
P2 = 15:23:27.8 UT1	U2 = 14:10:14.9 UT1
P3 = 16:36:05.7 UT1	U3 = 17:49:06.6 UT1
P4 = 19:01:37.8 UT1	U4 = 17:55:24.3 UT1

Circumstances at Greatest Eclipse

Lat. = 31°18.2'S	Sun Alt. = 72.7°
Long. = 048°28.0'W	Sun Azm. = 333.5°
Path Width = 281.5 km	Duration = 07m50.9s

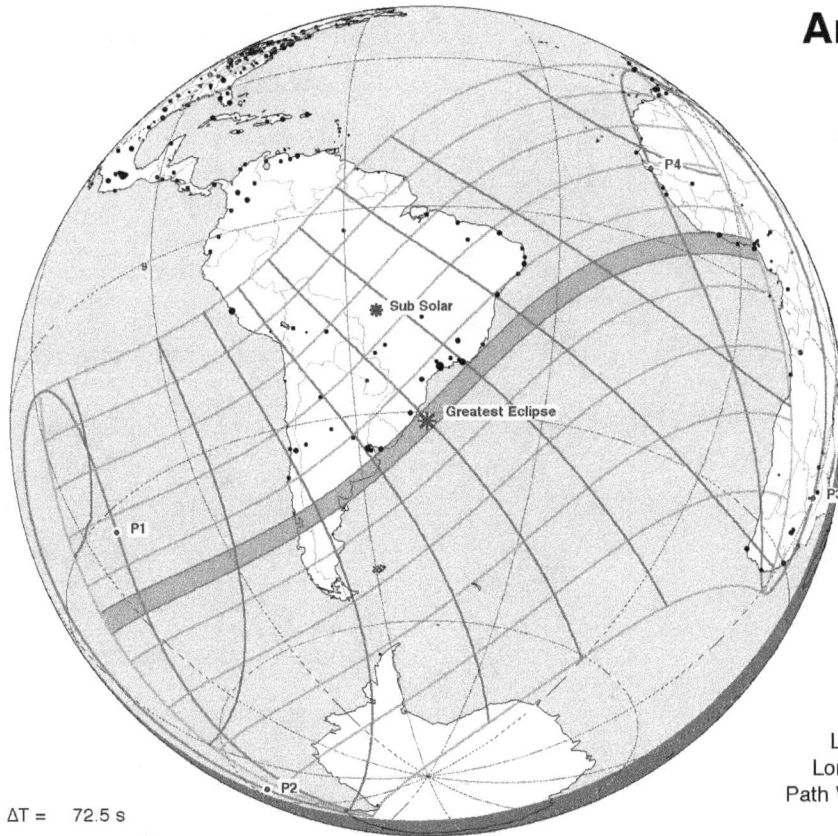

ΔT =   72.5 s

www.EclipseWise.com

©2020 F. Espenak

Total Solar Eclipse
2027 Aug 02

Greatest Eclipse = 10:06:37.4 UT1

Eclipse Magnitude = 1.0790

Eclipse Obscuration = 1.1643

Gamma = 0.1421

Saros = 136   [38 of 71]

Decending Node

Geocentric Coordinates at Greatest Eclipse

Sun	Moon
R.A. = 08h49m26.9s	R.A. = 08h49m40.1s
Dec. = +17°45'41.3"	Dec. = +17°53'47.8"
S.D. = 00°15'45.5"	S.D. = 00°16'43.1"
H.P. = 00°00'08.7"	H.P. = 01°01'21.4"

External/Internal Contacts of Penumbra	External/Internal Contacts of Umbra
P1 = 07:30:09.1 UT1	U1 = 08:23:25.0 UT1
P2 = 09:20:48.2 UT1	U2 = 08:26:38.4 UT1
P3 = 10:52:34.2 UT1	U3 = 11:46:40.4 UT1
P4 = 12:43:08.5 UT1	U4 = 11:49:53.1 UT1

Circumstances at Greatest Eclipse

Lat. = 25°30.2'N	Sun Alt. = 81.7°
Long. = 033°11.0'E	Sun Azm. = 202.0°
Path Width = 257.7 km	Duration = 06m22.6s

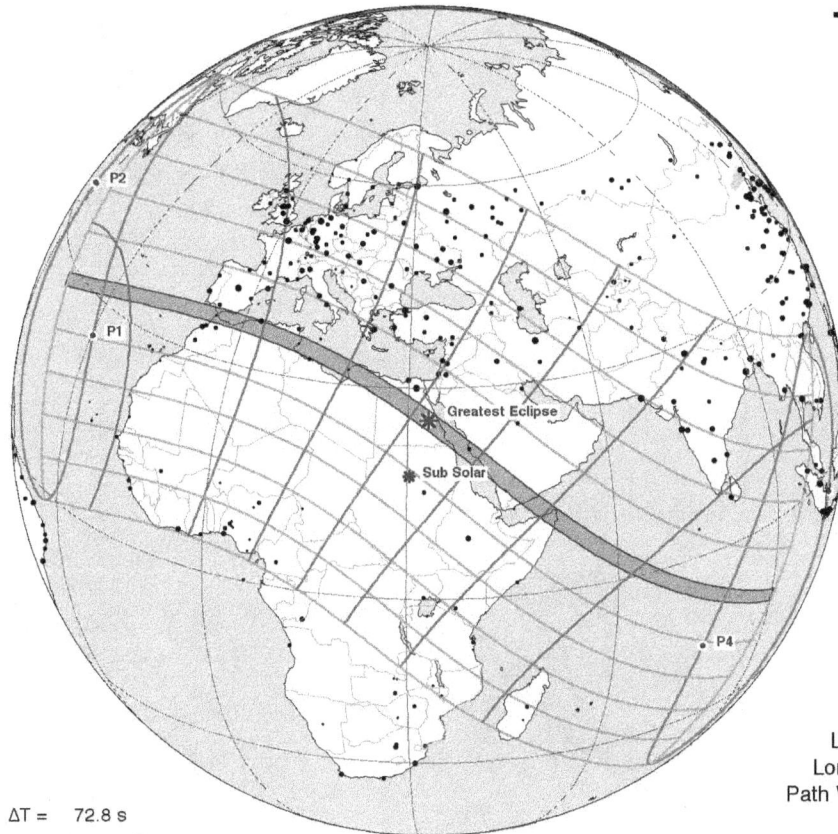

ΔT =   72.8 s

www.EclipseWise.com

©2020 F. Espenak

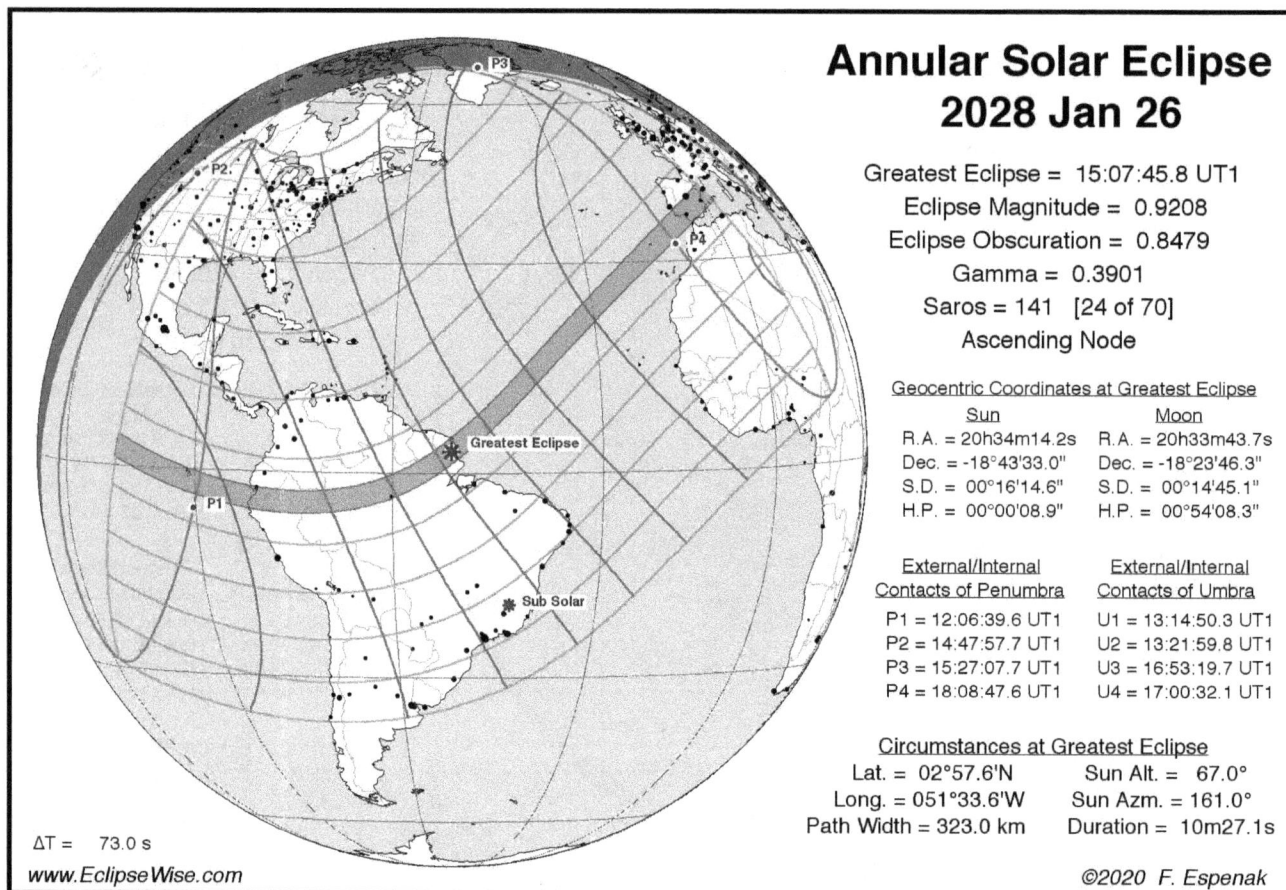

Annular Solar Eclipse
2028 Jan 26

Greatest Eclipse = 15:07:45.8 UT1
Eclipse Magnitude = 0.9208
Eclipse Obscuration = 0.8479
Gamma = 0.3901
Saros = 141 [24 of 70]
Ascending Node

Geocentric Coordinates at Greatest Eclipse

Sun	Moon
R.A. = 20h34m14.2s	R.A. = 20h33m43.7s
Dec. = -18°43'33.0"	Dec. = -18°23'46.3"
S.D. = 00°16'14.6"	S.D. = 00°14'45.1"
H.P. = 00°00'08.9"	H.P. = 00°54'08.3"

External/Internal Contacts of Penumbra	External/Internal Contacts of Umbra
P1 = 12:06:39.6 UT1	U1 = 13:14:50.3 UT1
P2 = 14:47:57.7 UT1	U2 = 13:21:59.8 UT1
P3 = 15:27:07.7 UT1	U3 = 16:53:19.7 UT1
P4 = 18:08:47.6 UT1	U4 = 17:00:32.1 UT1

Circumstances at Greatest Eclipse

Lat. = 02°57.6'N	Sun Alt. = 67.0°
Long. = 051°33.6'W	Sun Azm. = 161.0°
Path Width = 323.0 km	Duration = 10m27.1s

ΔT = 73.0 s

www.EclipseWise.com

©2020 F. Espenak

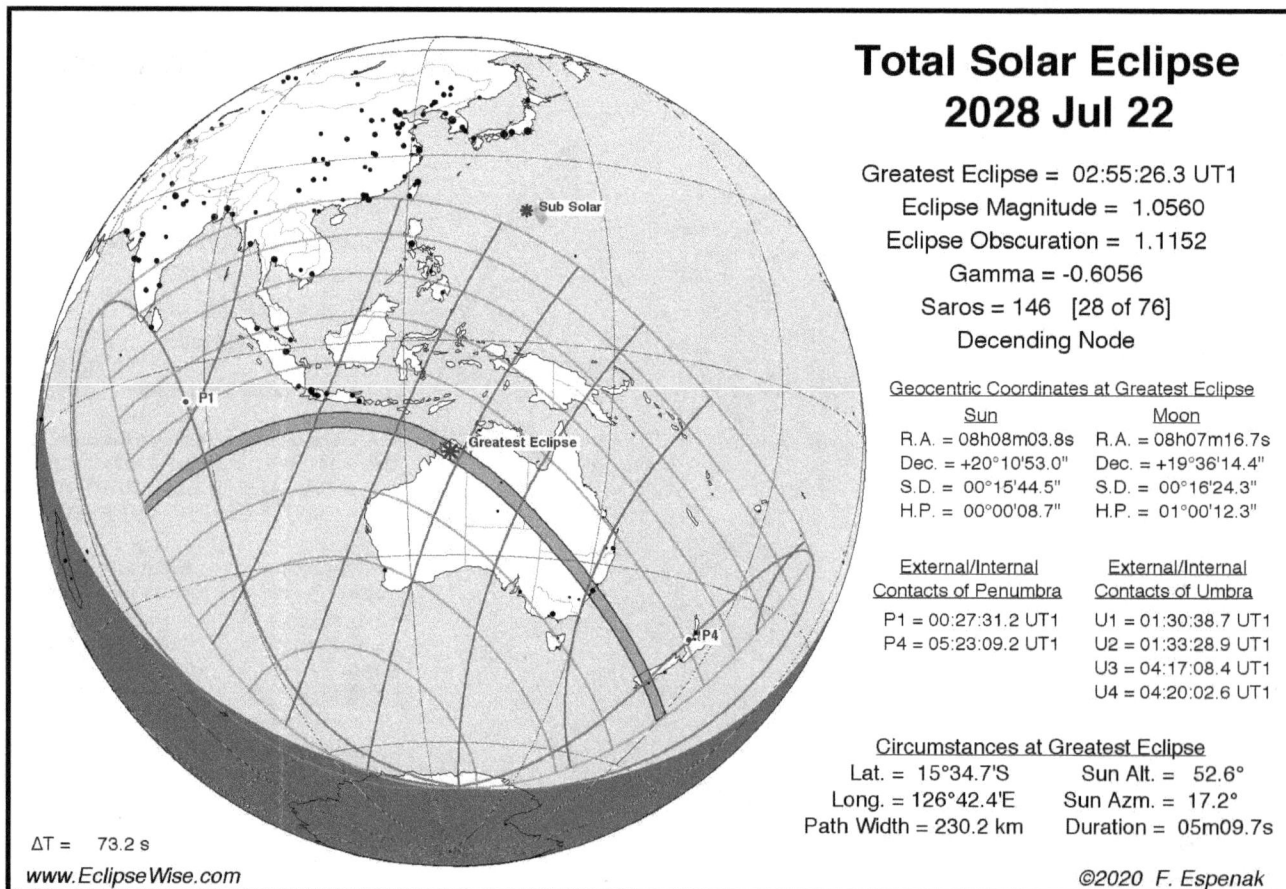

Total Solar Eclipse
2028 Jul 22

Greatest Eclipse = 02:55:26.3 UT1
Eclipse Magnitude = 1.0560
Eclipse Obscuration = 1.1152
Gamma = -0.6056
Saros = 146 [28 of 76]
Decending Node

Geocentric Coordinates at Greatest Eclipse

Sun	Moon
R.A. = 08h08m03.8s	R.A. = 08h07m16.7s
Dec. = +20°10'53.0"	Dec. = +19°36'14.4"
S.D. = 00°15'44.5"	S.D. = 00°16'24.3"
H.P. = 00°00'08.7"	H.P. = 01°00'12.3"

External/Internal Contacts of Penumbra	External/Internal Contacts of Umbra
P1 = 00:27:31.2 UT1	U1 = 01:30:38.7 UT1
P4 = 05:23:09.2 UT1	U2 = 01:33:28.9 UT1
	U3 = 04:17:08.4 UT1
	U4 = 04:20:02.6 UT1

Circumstances at Greatest Eclipse

Lat. = 15°34.7'S	Sun Alt. = 52.6°
Long. = 126°42.4'E	Sun Azm. = 17.2°
Path Width = 230.2 km	Duration = 05m09.7s

ΔT = 73.2 s

www.EclipseWise.com

©2020 F. Espenak

Partial Solar Eclipse
2029 Jan 14

Greatest Eclipse = 17:12:34.1 UT1
Eclipse Magnitude = 0.8714
Eclipse Obscuration = 0.8160
Gamma = 1.0553
Saros = 151 [15 of 72]
Ascending Node

Geocentric Coordinates at Greatest Eclipse

	Sun	Moon
R.A. =	19h47m03.1s	19h45m53.5s
Dec. =	-21°09'31.8"	-20°12'32.3"
S.D. =	00°16'15.6"	00°15'20.6"
H.P. =	00°00'08.9"	00°56'18.7"

External/Internal
Contacts of Penumbra
P1 = 15:01:55.5 UT1
P4 = 19:23:04.2 UT1

Circumstances at Greatest Eclipse
Lat. = 63°42.1'N   Sun Alt. = 0.0°
Long. = 114°13.5'W   Sun Azm. = 144.6°

ΔT = 73.4 s
www.EclipseWise.com   ©2020 F. Espenak

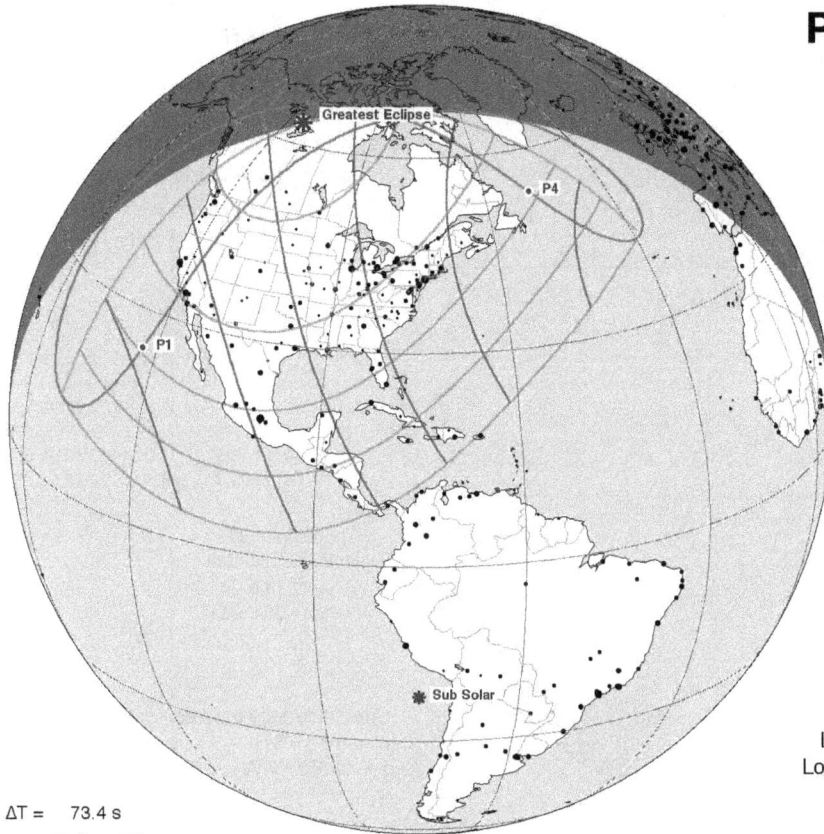

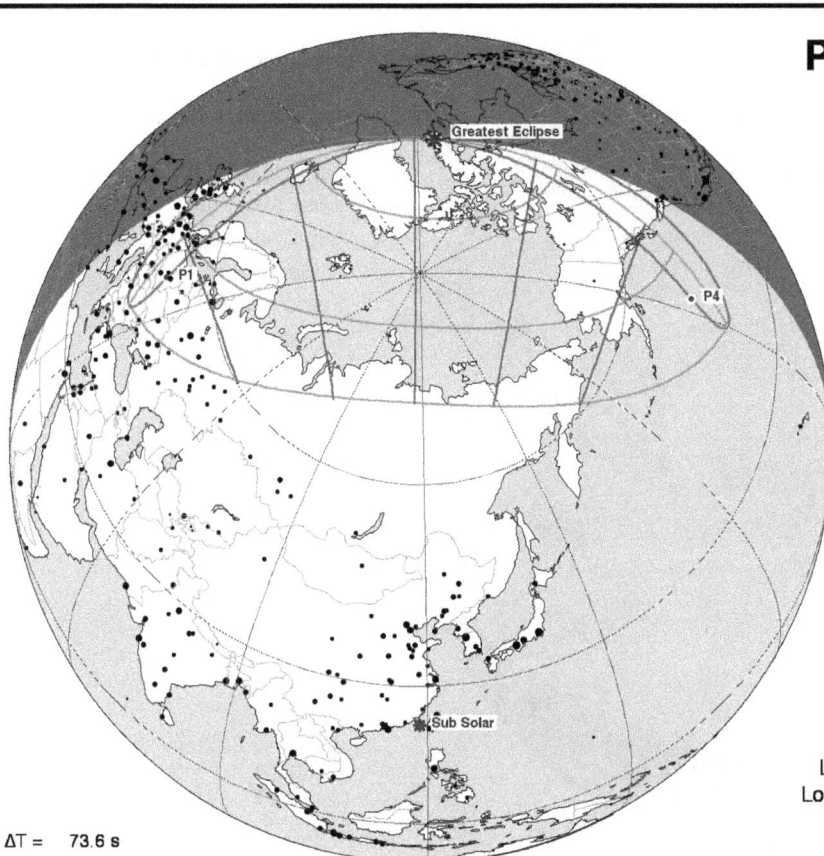

Partial Solar Eclipse
2029 Jun 12

Greatest Eclipse = 04:04:59.4 UT1
Eclipse Magnitude = 0.4576
Eclipse Obscuration = 0.3411
Gamma = 1.2943
Saros = 118 [69 of 72]
Decending Node

Geocentric Coordinates at Greatest Eclipse

	Sun	Moon
R.A. =	05h22m58.2s	05h23m08.9s
Dec. =	+23°09'45.7"	+24°21'37.7"
S.D. =	00°15'45.0"	00°15'10.6"
H.P. =	00°00'08.7"	00°55'42.0"

External/Internal
Contacts of Penumbra
P1 = 02:26:27.0 UT1
P4 = 05:43:29.1 UT1

Circumstances at Greatest Eclipse
Lat. = 66°45.9'N   Sun Alt. = 0.0°
Long. = 066°11.0'W   Sun Azm. = 355.5°

ΔT = 73.6 s
www.EclipseWise.com   ©2020 F. Espenak

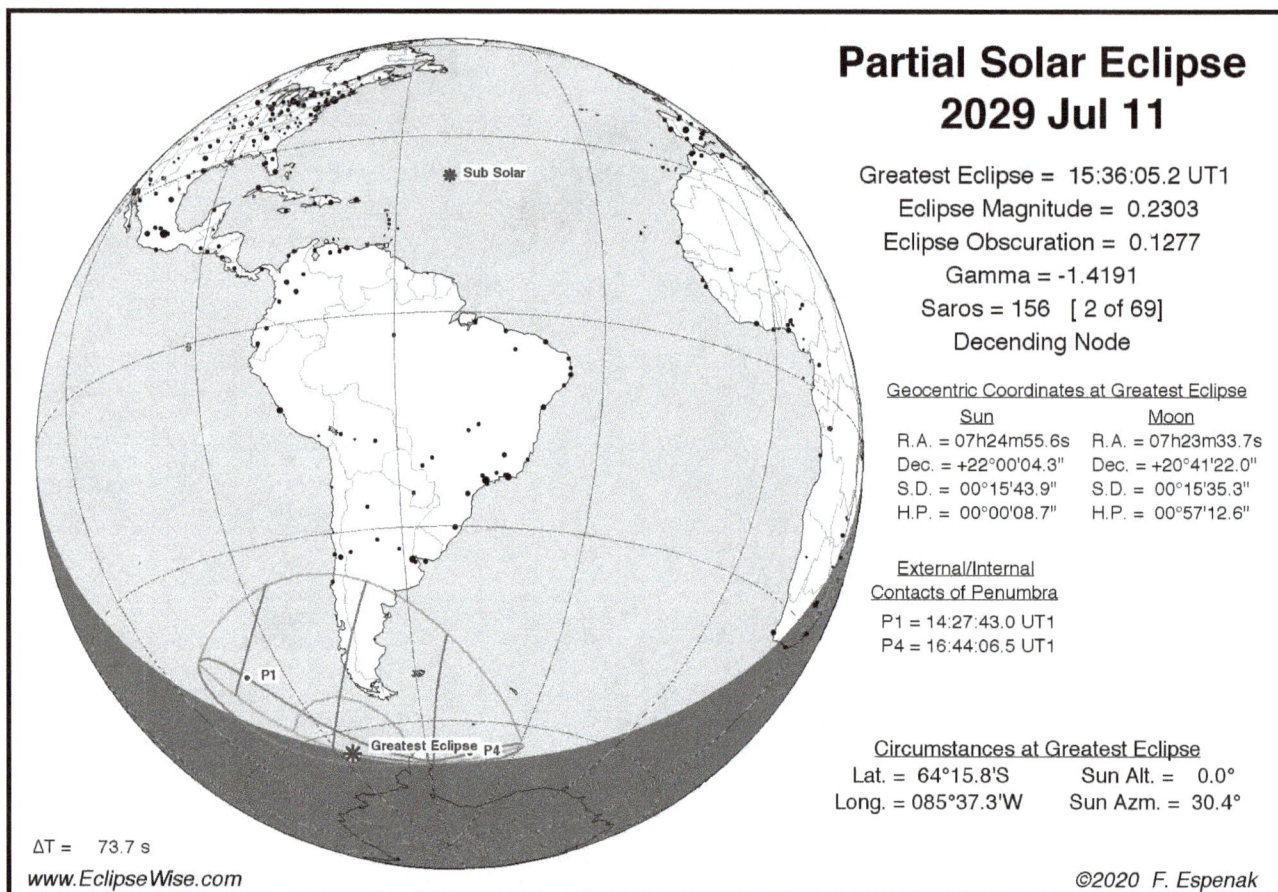

Partial Solar Eclipse
2029 Jul 11

Greatest Eclipse = 15:36:05.2 UT1

Eclipse Magnitude = 0.2303

Eclipse Obscuration = 0.1277

Gamma = -1.4191

Saros = 156 [2 of 69]

Decending Node

Geocentric Coordinates at Greatest Eclipse

	Sun	Moon
R.A. =	07h24m55.6s	R.A. = 07h23m33.7s
Dec. =	+22°00'04.3"	Dec. = +20°41'22.0"
S.D. =	00°15'43.9"	S.D. = 00°15'35.3"
H.P. =	00°00'08.7"	H.P. = 00°57'12.6"

External/Internal
Contacts of Penumbra

P1 = 14:27:43.0 UT1

P4 = 16:44:06.5 UT1

Circumstances at Greatest Eclipse

Lat. = 64°15.8'S	Sun Alt. = 0.0°
Long. = 085°37.3'W	Sun Azm. = 30.4°

ΔT = 73.7 s

www.EclipseWise.com

©2020 F. Espenak

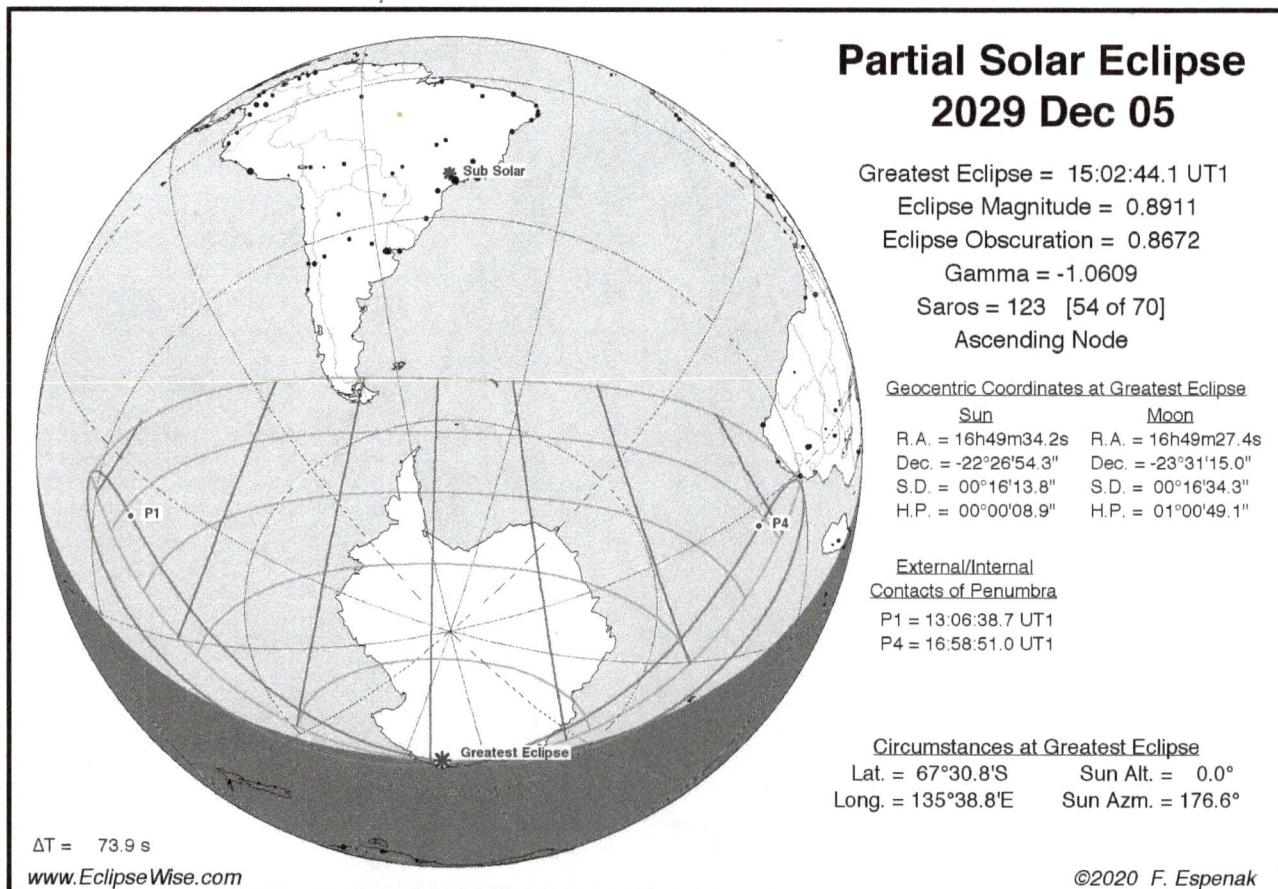

Partial Solar Eclipse
2029 Dec 05

Greatest Eclipse = 15:02:44.1 UT1

Eclipse Magnitude = 0.8911

Eclipse Obscuration = 0.8672

Gamma = -1.0609

Saros = 123 [54 of 70]

Ascending Node

Geocentric Coordinates at Greatest Eclipse

	Sun	Moon
R.A. =	16h49m34.2s	R.A. = 16h49m27.4s
Dec. =	-22°26'54.3"	Dec. = -23°31'15.0"
S.D. =	00°16'13.8"	S.D. = 00°16'34.3"
H.P. =	00°00'08.9"	H.P. = 01°00'49.1"

External/Internal
Contacts of Penumbra

P1 = 13:06:38.7 UT1

P4 = 16:58:51.0 UT1

Circumstances at Greatest Eclipse

Lat. = 67°30.8'S	Sun Alt. = 0.0°
Long. = 135°38.8'E	Sun Azm. = 176.6°

ΔT = 73.9 s

www.EclipseWise.com

©2020 F. Espenak

Annular Solar Eclipse
2030 Jun 01

Greatest Eclipse = 06:27:58.7 UT1

Eclipse Magnitude = 0.9443

Eclipse Obscuration = 0.8916

Gamma = 0.5626

Saros = 128 [59 of 73]

Decending Node

Geocentric Coordinates at Greatest Eclipse

Sun	Moon
R.A. = 04h37m01.2s	R.A. = 04h36m55.8s
Dec. = +22°03'55.3"	Dec. = +22°34'11.5"
S.D. = 00°15'46.4"	S.D. = 00°14'42.7"
H.P. = 00°00'08.7"	H.P. = 00°53'59.6"

External/Internal Contacts of Penumbra	External/Internal Contacts of Umbra
P1 = 03:34:39.2 UT1	U1 = 04:47:11.7 UT1
P4 = 09:21:15.7 UT1	U2 = 04:52:55.3 UT1
	U3 = 08:03:00.7 UT1
	U4 = 08:08:43.2 UT1

Circumstances at Greatest Eclipse

Lat. = 56°31.5'N	Sun Alt. = 55.5°
Long. = 080°03.7'E	Sun Azm. = 176.1°
Path Width = 249.6 km	Duration = 05m20.8s

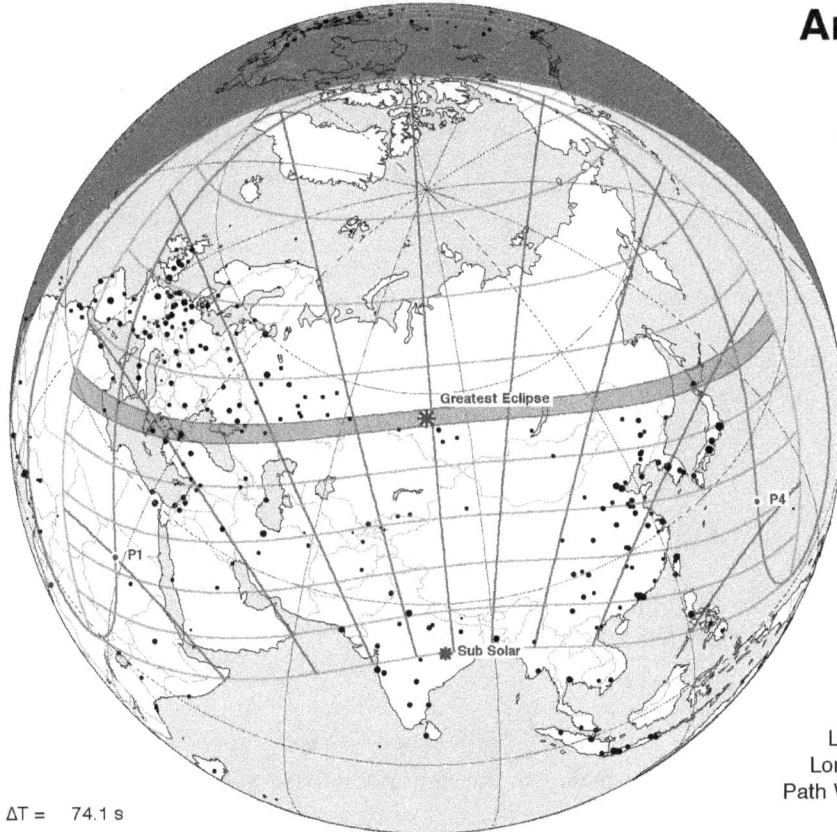

ΔT = 74.1 s

www.EclipseWise.com

©2020 F. Espenak

Total Solar Eclipse
2030 Nov 25

Greatest Eclipse = 06:50:22.5 UT1

Eclipse Magnitude = 1.0468

Eclipse Obscuration = 1.0959

Gamma = -0.3867

Saros = 133 [46 of 72]

Ascending Node

Geocentric Coordinates at Greatest Eclipse

Sun	Moon
R.A. = 16h03m58.7s	R.A. = 16h03m49.1s
Dec. = -20°45'39.0"	Dec. = -21°09'10.6"
S.D. = 00°16'12.1"	S.D. = 00°16'41.7"
H.P. = 00°00'08.9"	H.P. = 01°01'16.4"

External/Internal Contacts of Penumbra	External/Internal Contacts of Umbra
P1 = 04:16:41.1 UT1	U1 = 05:14:19.8 UT1
P2 = 06:24:40.3 UT1	U2 = 05:16:05.4 UT1
P3 = 07:15:59.3 UT1	U3 = 08:24:36.1 UT1
P4 = 09:24:00.7 UT1	U4 = 08:26:23.2 UT1

Circumstances at Greatest Eclipse

Lat. = 43°36.6'S	Sun Alt. = 67.0°
Long. = 071°13.7'E	Sun Azm. = 7.0°
Path Width = 169.3 km	Duration = 03m43.5s

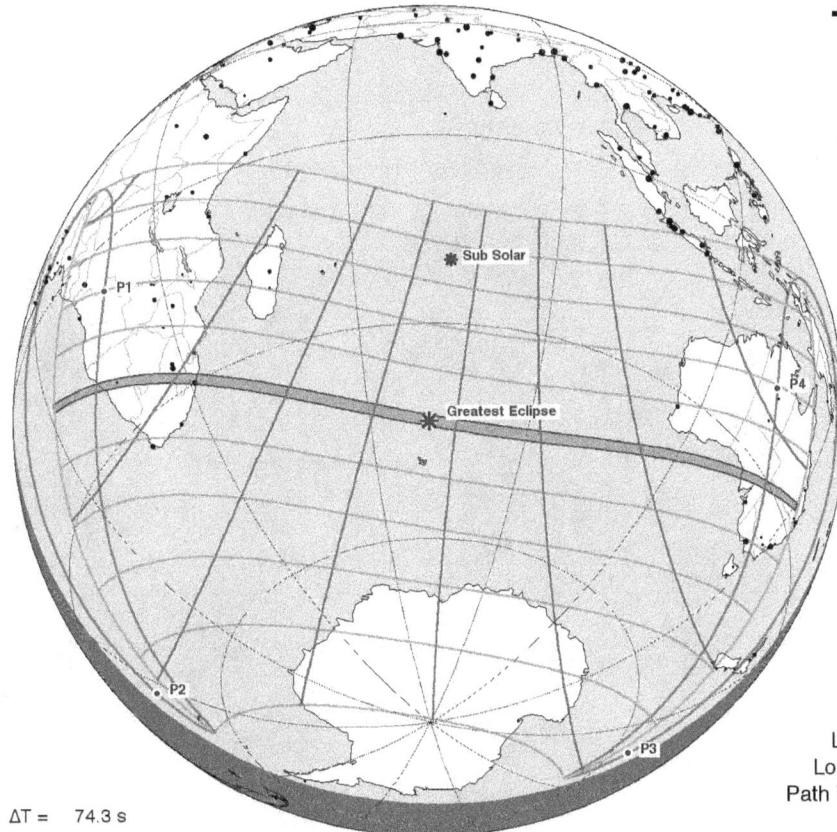

ΔT = 74.3 s

www.EclipseWise.com

©2020 F. Espenak

Key to Catalog of Solar Eclipses

The following catalog lists all solar eclipses for a 25-year period.

A brief description of each parameter in the catalog appears below.

Date — Gregorian date of Greatest Eclipse

Greatest Eclipse — Universal Time of Greatest Eclipse

Saros Num — Saros Series Number of the eclipse

Ecl Type — Solar Eclipse Type

      P = Partial Solar Eclipse
      A = Annular Solar Eclipse
      T = Total Solar Eclipse
      H = Hybrid Solar Eclipse

      + = Non-central eclipse of with no northern limit
      – = Non-central eclipse of with no southern limit

      b = Saros series begins (first eclipse in a Saros series)
      e = Saros series ends (last eclipse in a Saros series)

Gamma — minimum distance from the axis of the lunar shadow to the center of Earth

Mag — Eclipse Magnitude; fraction of the Sun's diameter obscured by the Moon

Lat & Long — latitude and longitude where the Sun appears in the zenith at greatest eclipse

Alt — altitude of the Sun at greatest eclipse

Duration — Central Line Duration (minutes. seconds) at greatest eclipse

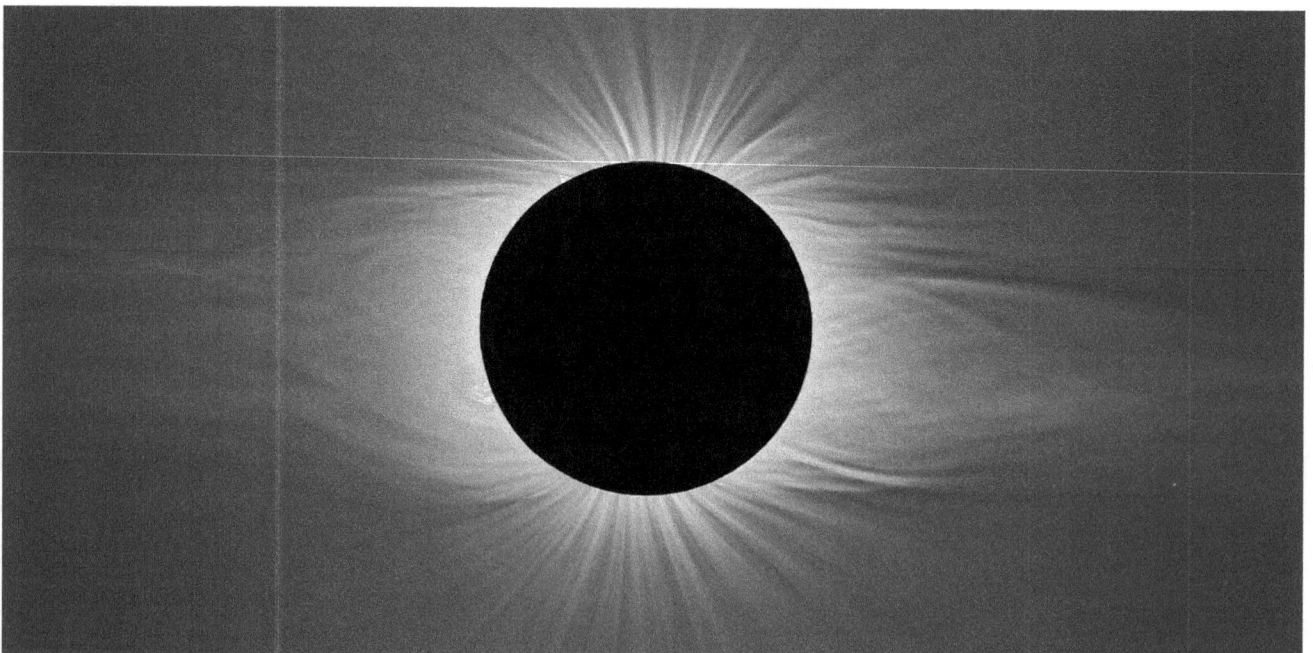

Photo 1–8 The solar corona is revealed during totality. Total Solar Eclipse of 2018 Jul 03. ©2018 F. Espenak

Catalog of Solar Eclipses: 2021 to 2045

Date	Greatest Eclipse	Saros	Type	Gamma	Mag	Lat	Long	Alt	Duration
2021 Jun 10	10:41:57	147	A	0.9152	0.9435	80.8N	66.8W	23	03m51s
2021 Dec 04	7:33:28	152	T	-0.9526	1.0367	76.8S	46.2W	17	01m54s
2022 Apr 30	20:41:26	119	P	-1.1901	0.6396	62.1S	71.5W	0	
2022 Oct 25	11:00:09	124	P	1.0701	0.8619	61.6N	77.3E	0	
2023 Apr 20	4:16:45	129	H	-0.3952	1.0132	9.6S	125.8E	67	01m16s
2023 Oct 14	17:59:30	134	A	0.3753	0.9520	11.4N	83.1W	68	05m17s
2024 Apr 08	18:17:18	139	T	0.3431	1.0566	25.3N	104.1W	70	04m28s
2024 Oct 02	18:45:02	144	A	-0.3509	0.9326	22.0S	114.5W	69	07m25s
2025 Mar 29	10:47:24	149	P	1.0405	0.9376	61.1N	77.1W	0	
2025 Sep 21	19:41:52	154	P	-1.0651	0.8550	60.9S	153.5E	0	
2026 Feb 17	12:11:54	121	A	-0.9743	0.9630	64.7S	86.7E	12	02m20s
2026 Aug 12	17:45:53	126	T	0.8977	1.0386	65.2N	25.2W	26	02m18s
2027 Feb 06	15:59:35	131	A	-0.2952	0.9281	31.3S	48.5W	73	07m51s
2027 Aug 02	10:06:37	136	T	0.1421	1.0790	25.5N	33.2E	82	06m23s
2028 Jan 26	15:07:46	141	A	0.3901	0.9208	3.0N	51.6W	67	10m27s
2028 Jul 22	2:55:26	146	T	-0.6056	1.0560	15.6S	126.7E	53	05m10s
2029 Jan 14	17:12:34	151	P	1.0553	0.8714	63.7N	114.2W	0	
2029 Jun 12	4:04:59	118	P	1.2943	0.4576	66.8N	66.2W	0	
2029 Jul 11	15:36:05	156	P	-1.4191	0.2303	64.3S	85.6W	0	
2029 Dec 05	15:02:44	123	P	-1.0609	0.8911	67.5S	135.6E	0	
2030 Jun 01	6:27:59	128	A	0.5626	0.9443	56.5N	80.1E	55	05m21s
2030 Nov 25	6:50:23	133	T	-0.3867	1.0468	43.6S	71.2E	67	03m44s
2031 May 21	7:14:50	138	A	-0.1970	0.9589	8.9N	71.7E	79	05m26s
2031 Nov 14	21:06:16	143	H	0.3078	1.0106	0.6S	137.6W	72	01m08s
2032 May 09	13:25:27	148	A	-0.9375	0.9957	51.3S	7.1W	20	00m22s
2032 Nov 03	5:32:58	153	P	1.0643	0.8554	70.4N	132.6E	0	
2033 Mar 30	18:01:20	120	T	0.9778	1.0462	71.3N	155.8W	11	02m37s
2033 Sep 23	13:53:15	125	P	-1.1583	0.6890	72.2S	121.3W	0	
2034 Mar 20	10:17:29	130	T	0.2894	1.0458	16.1N	22.2E	73	04m09s
2034 Sep 12	16:18:11	135	A	-0.3936	0.9736	18.2S	72.6W	67	02m58s
2035 Mar 09	23:04:37	140	A	-0.4368	0.9919	29.0S	155.0W	64	00m48s
2035 Sep 02	1:55:30	145	T	0.3727	1.0320	29.1N	158.0E	68	02m54s
2036 Feb 27	4:45:32	150	P	-1.1942	0.6286	71.6S	131.5W	0	
2036 Jul 23	10:30:49	117	P	-1.4250	0.1992	68.9S	3.5E	0	
2036 Aug 21	17:24:28	155	P	1.0825	0.8623	71.1N	47.0E	0	
2037 Jan 16	9:47:38	122	P	1.1477	0.7049	68.5N	20.8E	0	
2037 Jul 13	2:39:18	127	T	-0.7246	1.0413	24.8S	139.1E	43	03m58s
2038 Jan 05	13:45:53	132	A	0.4169	0.9728	2.1N	25.5W	65	03m18s
2038 Jul 02	13:31:37	137	A	0.0398	0.9911	25.4N	21.9W	88	01m00s
2038 Dec 26	0:58:51	142	T	-0.2881	1.0269	40.3S	163.9E	73	02m18s
2039 Jun 21	17:11:35	147	A	0.8312	0.9454	78.9N	102.1W	33	04m05s
2039 Dec 15	16:22:27	152	T	-0.9458	1.0356	80.9S	172.8E	18	01m51s
2040 May 11	3:41:43	119	P	-1.2529	0.5306	62.8S	174.4E	0	
2040 Nov 04	19:07:42	124	P	1.0993	0.8074	62.2N	53.4W	0	
2041 Apr 30	11:51:01	129	T	-0.4492	1.0189	9.6S	12.2E	63	01m51s
2041 Oct 25	1:35:02	134	A	0.4133	0.9467	9.9N	162.8E	66	06m07s
2042 Apr 20	2:16:10	139	T	0.2956	1.0614	27.0N	137.3E	73	04m51s
2042 Oct 14	1:59:21	144	A	-0.3030	0.9301	23.7S	137.8E	72	07m44s
2043 Apr 09	18:56:28	149	T+	1.0031	1.0096	61.3N	151.9E	0	01m46s
2043 Oct 03	3:00:28	154	A-	-1.0102	0.9497	61.0S	35.2E	0	02m17s
2044 Feb 28	20:23:18	121	As	-0.9954	0.9600	62.2N	25.6W	4	02m27s
2044 Aug 23	1:15:40	126	T	0.9613	1.0364	64.3N	120.5W	15	02m04s
2045 Feb 16	23:54:45	131	A	-0.3125	0.9285	28.3S	166.2W	72	07m47s
2045 Aug 12	17:41:17	136	T	0.2116	1.0774	25.9N	78.6W	78	06m06s

Photo 2–1 Phases of the total lunar eclipse of 2014 Apr 15 are captured in this multiple exposure sequence. ©2014 F. Espenak

Section 2: Lunar Eclipses

Introduction

The Moon orbits Earth once every 29.5306 days with respect to the Sun. Over the course of its orbit, the Moon's changing position relative to the Sun results in its familiar phases: New Moon > First Quarter > Full Moon > Last Quarter > New Moon. The New Moon phase is the only one not visible because the illuminated side of the Moon points away from Earth.

During Full Moon, the Moon appears opposite the Sun in the sky. It rises as the Sun sets and is visible throughout the night. The Full Moon sets in the morning just as the Sun rises. This geometry occurs when the Moon is 180° from the Sun as seen from Earth. It corresponds to the direction Earth casts its shadow into space.

The Moon's orbit is tilted about 5.1° to Earth's orbit around the Sun. As seen from Earth, the points where the two orbits appear to cross are called the nodes. When the Full Moon occurs near one of these nodes, the Moon can pass through some portion of Earth's shadow and a lunar eclipse occurs.

Earth's shadow has two cone-shaped components, one nested inside the other. The outer or penumbral shadow is a zone where the Sun's rays are partially blocked., The inner or umbral shadow is a region where direct rays from the Sun are completely blocked.

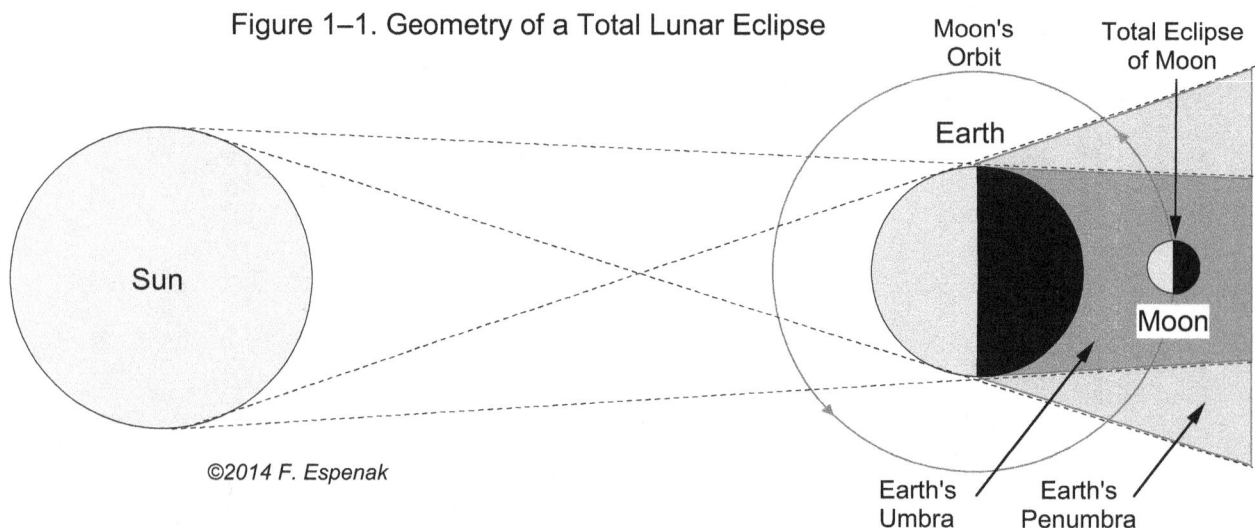

Figure 2–1 illustrates the geometry of a total lunar eclipse. A partial eclipse is visible if only part of the Moon enters Earth's umbral shadow. If the Moon passes through the penumbral shadow but misses the umbral shadow, then a penumbral eclipse occurs.

Photo 2–2 Examples of the visual appearance of a penumbral, partial, and total lunar eclipse. ©2020 F. Espenak

Types of Lunar Eclipses

There are three types of lunar eclipses:

1. **Penumbral Lunar Eclipse** — The Moon passes through Earth's faint penumbral shadow. Penumbral eclipses are of minor interest since they are quite subtle and difficult to observe.
2. **Partial Lunar Eclipse** — A portion of the Moon passes through Earth's dark umbral shadow. The remaining part of the Moon appears bright even though it lies deep within the penumbra. Partial eclipses are easy to see, even with the unaided eye.
3. **Total Lunar Eclipse** — The entire Moon passes through Earth's umbral shadow. Total eclipses are quite striking for the vibrant range of color of the Moon during the total phase, referred to as totality.

Figure 1–2. Types of Lunar Eclipses

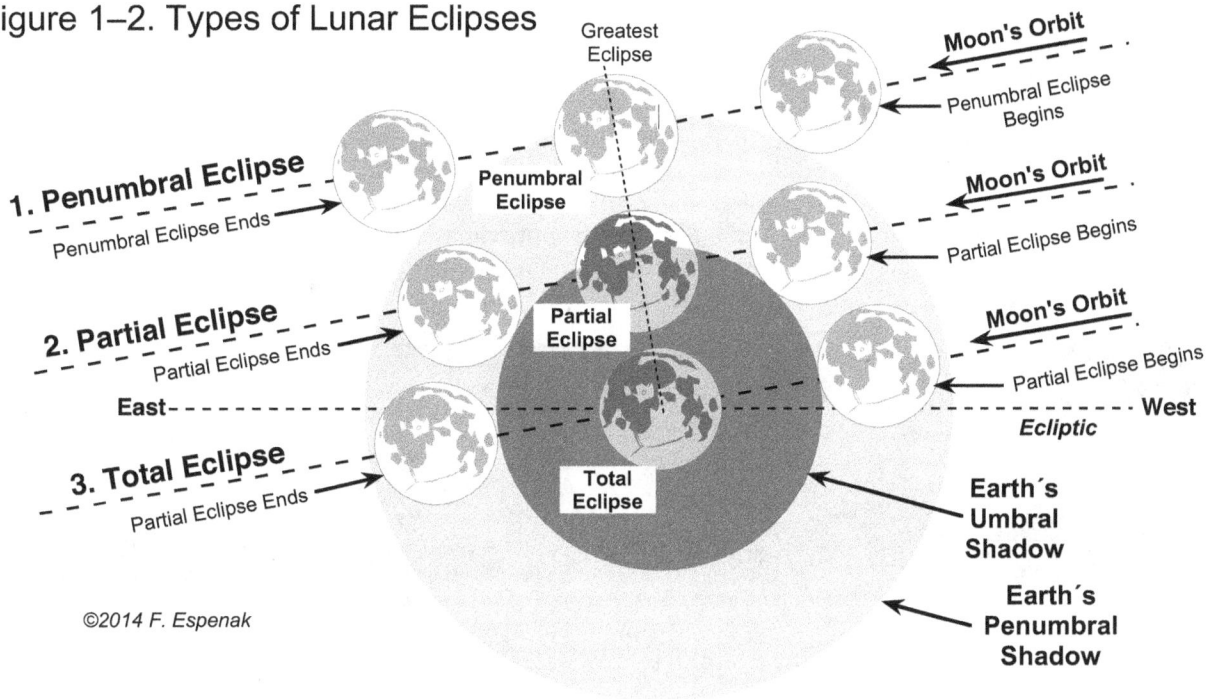

Figure 2–2 illustrates the three types of lunar eclipses as seen from Earth. 1) A penumbral eclipse occurs when the Moon passes through the penumbra but completely misses the umbra. 2) A partial eclipse occurs if some portion of the Moon enters the umbra. 3) A total lunar eclipse takes place when the entire disk of the Moon enters the umbral shadow.

Photo 2–3 Penumbral lunar eclipse of 2002 Nov 20. Left: Moon before the eclipse begins. Right: mid-eclipse when 88.9% of the Moon's is in the penumbral shadow. ©2002 F. Espenak

Visual Appearance of Penumbral Lunar Eclipses

The visual appearance of penumbral, partial and total lunar eclipses differs significantly. While penumbral eclipses are pale and difficult to see, partial eclipses are easy naked-eye events while total eclipses are colorful and dramatic.

Earth's penumbral shadow forms a diverging cone that expands into space away from the Sun. Within this zone, Earth blocks part but not all of the Sun's disk. Some portion of the Sun's direct rays continue to reach the Moon during a penumbral eclipse.

The early and late stages of a penumbral eclipse are completely invisible to the eye. It is only when about 2/3 of the Moon's disk has entered the penumbral shadow that a skilled observer can detect a faint shading across the Moon.

Even when 90% of the Moon is immersed in the penumbra, approximately 10% of the Sun's rays still reach the Moon's deepest limb. Under such conditions, the Moon remains relatively bright with only a subtle shadow gradient across its disk.

Photo 2–4 Time sequence of the partial lunar eclipse of 2012 Jun 04. ©2012 F. Espenak

Visual Appearance of Partial Lunar Eclipses

Compared to penumbral eclipses, partial eclipses are easy to see with the naked eye. The lunar limb extending into the umbral shadow appears very dark or even black. This is due to a contrast effect since the remaining portion of the Moon in the penumbra is hundreds of times brighter. Because the umbral shadow's diameter is about 2.7 times the Moon's diameter, it appears as though a semi-circular bite has been taken out of the Moon.

Aristotle (384–322 BCE) first proved that Earth was round using the curved umbral shadow seen at partial eclipses. In comparing observations of several eclipses, he noted that Earth's shadow was round no matter where the eclipse took place, whether the Moon was high in the sky or low near the horizon. Aristotle reasoned that only a sphere casts a round shadow from every angle.

Photo 2–5 The beginning, middle and end of totality during the total lunar eclipse of 2004 October 28. ©2004 F. Espenak

Visual Appearance of Total Lunar Eclipses

A total lunar eclipse is the most dramatic and visually compelling type of lunar eclipse. The Moon's appearance can vary enormously throughout the period of totality and from one eclipse to the next. The geometry of the Moon's path through the umbra plays a significant role in determining the appearance of totality. The effect that Earth's atmosphere has on a total eclipse is not as apparent. Although the physical mass of Earth blocks all direct sunlight from the umbra, the planet's atmosphere filters, attenuates and bends some of the Sun's rays into the shadow.

The molecules in Earth's atmosphere scatter short wavelength light (i.e., yellow, green, blue) more than long wavelength light (i.e., orange, red). The same process responsible for making sunsets red also gives total lunar eclipses their characteristic ruddy color. The exact appearance can vary widely in both hue and brightness.

Because the lowest layers of the atmosphere are the thickest, they absorb more sunlight and refract it through larger angles. About 75% of the atmosphere's mass is concentrated in the bottom 10 kilometers (troposphere) as well as most of the water vapor, which can form massive clouds that block even more light. Just above the troposphere lies the stratosphere (10 to 50 kilometers), a rarified zone above most of the planet's weather systems. The stratosphere is subject to important photochemical reactions due to the high level of solar ultraviolet radiation that penetrates the region. The troposphere and stratosphere act together as a ring-shaped lens that refracts heavily reddened sunlight into Earth's umbral shadow. Since the higher stratospheric layers contain less gas, they refract sunlight through progressively smaller angles into the outer parts of the umbra. In contrast, denser tropospheric layers refract sunlight through larger angles to reach the inner parts of the umbra.

Because of this lensing effect, the amount of light refracted into the umbra tends to decrease radially from the edge to the center. Inhomogeneities from varying amounts of cloud and dust at differing latitudes can cause significant variations in brightness throughout the umbra.

Besides water (cloud, mist, precipitation), Earth's atmosphere also contains aerosols or tiny particles of organic debris, meteoric dust, volcanic ash and photochemical droplets. This material attenuates sunlight before it is refracted into the umbra. For instance, major volcanic eruptions in 1963 (Agung) and 1982 (El Chichon) each dumped large quantities of gas and ash into the stratosphere and were followed by several years of dark eclipses.

The 1991 eruption of Pinatubo in the Philippines had a similar effect. While most of the solid ash fell to Earth several days after circulating through the troposphere, a sizable volume of sulfur dioxide (SO_2) reached the stratosphere where it interacts with water vapor and eventually produces sulfuric acid (H_2SO_4). This high-altitude volcanic haze layer severely attenuates sunlight that must travel several hundred kilometers horizontally through the layer before being refracted into the umbral shadow. Thus, total eclipses following large volcanic eruptions are particularly dark. The total lunar eclipse of 1992 Dec 09 (1½ years after Pinatubo) was so dark that it was difficult to see the Moon's dull gray disk with the naked eye.

All total eclipses begin with penumbral and partial phases. After the total phase, the eclipse ends with more partial and penumbral phases. Lunar eclipses are completely safe to view and require none of the precautions needed for viewing solar eclipses (like special filters). The best views of a lunar eclipse are with binoculars and the naked eye.

Figure 2–1. Lunar Eclipse Contacts

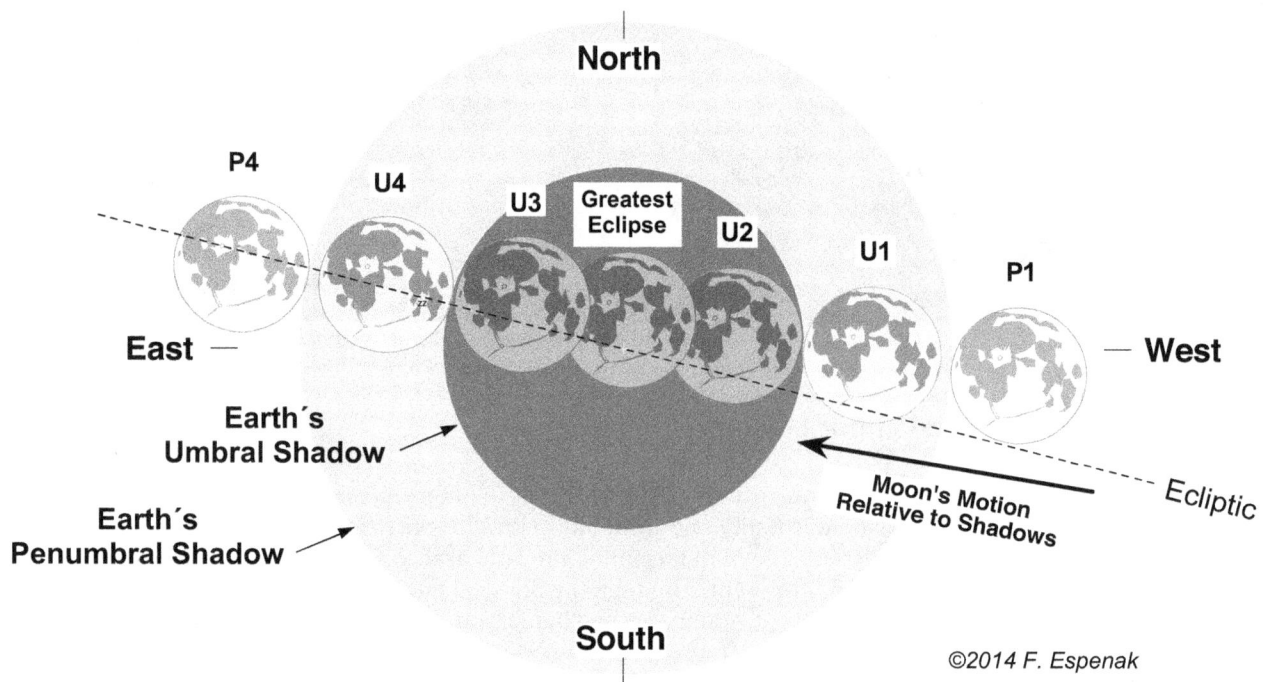

©2014 F. Espenak

Figure 2–3 illustrates the six contacts for a total lunar eclipse. These correspond to the instants when the Moon's disk is externally tangent to the penumbra (P1 and P4), or either externally or internally tangent to the umbra (U1, U2, U3, and U4). Partial eclipses do not have contacts U2 and U3, while penumbral eclipses only have contacts P1 and P4.

Lunar Eclipse Contacts

During the course of a lunar eclipse, the instants when the Moon's disk becomes tangent to Earth's shadows are known as eclipse contacts. They mark the primary stages or phases of a lunar eclipse (see figure 2-3) and are defined as follows.

P1 — Penumbral Eclipse Begins (Instant of first exterior tangency of the Moon with the Penumbra)
U1 — Partial Eclipse Begins (Instant of first exterior tangency of the Moon with the Umbra)
U2 — Total Eclipse Begins (Instant of first interior tangency of the Moon with the Umbra)
U3 — Total Eclipse Ends (Instant of last interior tangency of the Moon with the Umbra)
U4 — Partial Eclipse Ends (Instant of last exterior tangency of the Moon with the Umbra)
P4 — Penumbral Eclipse Ends (Instant of last exterior tangency of the Moon with the Penumbra)

Penumbral eclipses only have first and last contacts (i.e., P1 and P4) although neither of these events is observable.

In addition to P1 and P4, partial eclipses also have contacts U1 and U4 when the partial phases begin and end.

Total lunar eclipses have all six contacts. Contacts U2 and U3 mark the instants when the Moon's entire disk is first and last internally tangent to the umbra. These are the times when the total phase of the eclipse begins and ends..

The instant of greatest eclipse occurs when the Moon passes closest to the shadow axis. This is the maximum phase of the eclipse when the Moon is at its deepest position within either the penumbral or umbral shadow.

Photo 2–6 Five minutes before the start of totality (2018 Jan 31), the Moon is bathed in an orange-red light.
The narrow rim outside the umbra and still in sunlight appears brilliant white. ©2018 F. Espenak

Enlargement of Earth's Shadows

In 1707, Philippe de La Hire made a curious observation about Earth's umbra. The predicted radius of the shadow needed to be enlarged by about 1/41 in order to fit timings made during a lunar eclipse. The enlargement is attributed to Earth's atmosphere, which becomes increasingly less transparent at lower levels. Additional observations over the next two centuries revealed that the shadow enlargement was somewhat variable from one eclipse to the next.

William Chauvenet (1891) formulated one method to account for the shadow enlargement while André-Louis Danjon (1951) devised another. Both methods assumed a circular cross section for the umbral shadow. However, Earth is flattened at the poles and bulges at the Equator, so an oblate spheroid more closely represents its shape. The projection of each of the planet's shadows is an ellipse rather than a circle. Furthermore, Earth's axial tilt towards or away from the Sun throughout the year means the elliptical shape of the penumbral and umbral shadows varies as well.

In an analysis of 22,539 observations made at 94 lunar eclipses from 1842 to 2011, Herald and Sinnott[7] (2014) found that the size and shape of the umbra are consistent with an oblate spheroid at the time of each eclipse, enlarged by the empirically determined occulting layer that uniformly surrounds Earth. The effective height of this layer was found to be 87 kilometers. Based on this work, the authors developed a new method to calculate the shadow enlargement including their elliptical shape.

This new method is the most rigorous and accurate procedure to date. The *Eclipse Almanac* uses it in the lunar eclipse predictions presented here.

Explanation of Lunar Eclipse Figures

There are 22 eclipses of the Moon during the period 2021 to 2030. A figure for each eclipse is included.

Each figure consists of two diagrams. The first one depicts the Moon's path through Earth's penumbral and umbral shadows (with Celestial North up). The second is a map showing the geographic visibility of each eclipse phase. All features in these diagrams are identified in the key on the next page.

The Moon's orbital motion with respect to the shadows is from west to east (right to left). Each phase of the eclipse is defined by the instant when the Moon's limb is externally or internally tangent to the penumbra or umbra. The six primary contacts of the Moon with the penumbral and umbral shadows are defined as follows.

P1 — Penumbral Eclipse Begins   (Instant of first exterior tangency of the Moon with the Penumbra)
U1 — Partial Eclipse Begins   (Instant of first exterior tangency of the Moon with the Umbra)
U2 — Total Eclipse Begins   (Instant of first interior tangency of the Moon with the Umbra)
U3 — Total Eclipse Ends   (Instant of last interior tangency of the Moon with the Umbra)
U4 — Partial Eclipse Ends   (Instant of last exterior tangency of the Moon with the Umbra)
P4 — Penumbral Eclipse Ends   (Instant of last exterior tangency of the Moon with the Penumbra)

Penumbral lunar eclipses have two primary contacts: P1 and P4, but neither is observable because the edge of the penumbra is indistinct and extremely faint.

In addition to the penumbral contacts, partial lunar eclipses have two more contacts as the Moon's limb enters and exits the umbral shadow: U1 and U4, respectively. These two contacts mark the instants when the partial phase of the eclipse begins and ends.

Total lunar eclipses undergo all six contacts. The two additional umbral contacts are the instants when the Moon's entire disk is first and last internally tangent to the umbra: U2 and U3, respectively. They mark the times when the total phase of the eclipse begins and ends.

The Moon passes closest to the shadow axis at the instant of greatest eclipse. This corresponds to the maximum phase of the eclipse and the Moon's position at this instant is also depicted in the path diagrams.

The equidistant cylindrical projection map shows the geographic region of visibility at each phase of the eclipse. This is accomplished using a series of curves showing where Moonrise and Moonset occur at each eclipse contact. The map is shaded to indicate eclipse visibility. The entire eclipse is visible from the zone with no shading. Conversely, none of the eclipse can be seen from the zone with the darkest shading.

At greatest eclipse, the Moon is deepest in Earth's shadow. The geographic location where the Moon appears in the zenith at greatest eclipse is shown by an asterisk.

Parameters relevant to the eclipse appear in on the right side of each figure. The instant of greatest eclipse is the time (Universal Time[8] or UT1) when the Moon passes closest to the shadow axis. The penumbral and umbral magnitudes are the fractions of the Moon's diameter immersed in each shadow. Gamma is the minimum distance of the Moon's center from the axis of Earth's shadow at greatest eclipse. The Saros series of the eclipse, and the

[7] Herald, D., and Sinnott, R. W., "Analysis of Lunar Crater Timings, 1842–2011," *J. Br. Astron. Assoc.*, **124**, 5 (2014)

[8] Universal Time (UT1) is the modern-day replacement for Greenwich Mean Time and is based on Earth's rotation with respect to distant quasars.

node of the Moon's orbit are given. Each contact time of the Moon's edge with the penumbral and umbral shadows is listed in Universal Time (UT1). Depending on the eclipse type, the duration of the penumbral, partial or total phases are given.

2–7 A time sequence shows the partial phases and totality during the total lunar eclipse of 2000 July 16 from Maui, Hawaii (Copyright ©2000 by Fred Espenak).

Lunar Eclipse Figures

Key to Lunar Eclipse Figures

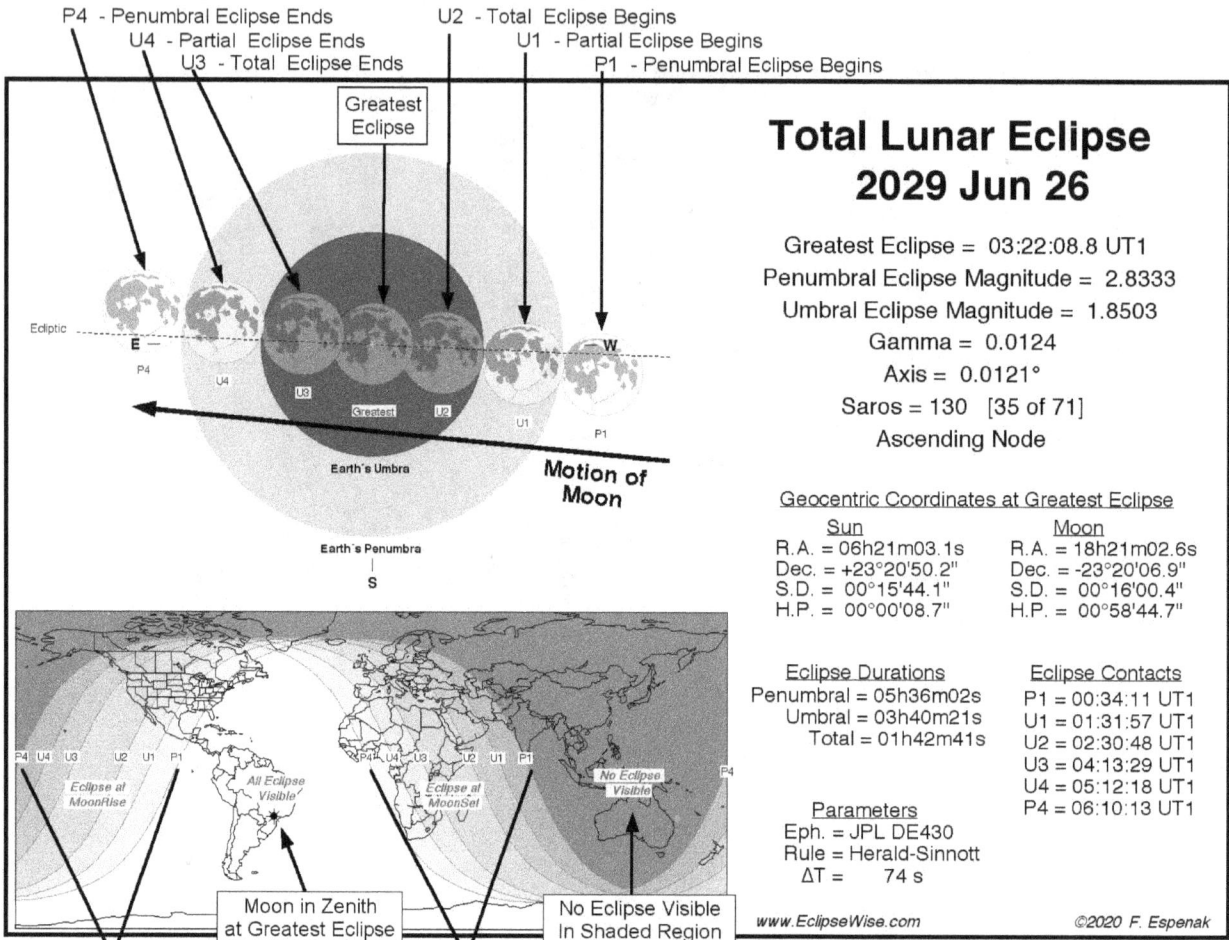

P4 - Penumbral Eclipse Ends
U4 - Partial Eclipse Ends
U3 - Total Eclipse Ends

U2 - Total Eclipse Begins
U1 - Partial Eclipse Begins
P1 - Penumbral Eclipse Begins

**Total Lunar Eclipse
2029 Jun 26**

Greatest Eclipse = 03:22:08.8 UT1
Penumbral Eclipse Magnitude = 2.8333
Umbral Eclipse Magnitude = 1.8503
Gamma = 0.0124
Axis = 0.0121°
Saros = 130 [35 of 71]
Ascending Node

Geocentric Coordinates at Greatest Eclipse

Sun	Moon
R.A. = 06h21m03.1s	R.A. = 18h21m02.6s
Dec. = +23°20'50.2"	Dec. = -23°20'06.9"
S.D. = 00°15'44.1"	S.D. = 00°16'00.4"
H.P. = 00°00'08.7"	H.P. = 00°58'44.7"

Eclipse Durations
Penumbral = 05h36m02s
Umbral = 03h40m21s
Total = 01h42m41s

Eclipse Contacts
P1 = 00:34:11 UT1
U1 = 01:31:57 UT1
U2 = 02:30:48 UT1
U3 = 04:13:29 UT1
U4 = 05:12:18 UT1
P4 = 06:10:13 UT1

Parameters
Eph. = JPL DE430
Rule = Herald-Sinnott
ΔT = 74 s

www.EclipseWise.com   ©2020 F. Espenak

Eclipse During Moon Rise
P1 - Penumbral Eclipse Begins
U1 - Partial Eclipse Begins
U2 - Total Eclipse Begins
U3 - Total Eclipse Ends
U4 - Partial Eclipse Ends
P4 - Penumbral Eclipse Ends

Eclipse During Moon Set
P1 - Penumbral Eclipse Begins
U1 - Partial Eclipse Begins
U2 - Total Eclipse Begins
U3 - Total Eclipse Ends
U4 - Partial Eclipse Ends
P4 - Penumbral Eclipse Ends

Layout of Lunar Eclipse Figure
Upper Left: Moon's Path Thru Earth's Shadows
Lower Left: Map of Eclipse Visibility
Right Side: Parameters for Lunar Eclipse

Explanation of Parameters Used in Lunar Eclipse Figures

Greatest Eclipse – The instant when the Moon passes closest to the axis of Earth's shadow cone (Universal Time[9])
Penumbral Eclipse Magnitude – Fraction of the Moon's diameter immersed in the penumbra at greatest eclipse.
Umbral Eclipse Magnitude – The fraction of the Moon's diameter immersed in the umbra at greatest eclipse.
Gamma – Minimum distance from the Moon's center to Earth's shadow axis (units of Earth's equatorial radius).
Axis – Minimum distance from the Moon's center to Earth's shadow axis (units of degrees).
Saros Series – The Saros series that the eclipse belongs to. The numbers in "[ ]" are the eclipse's sequential position and the number of eclipses in the Saros series.
Node – The orbital node near which the eclipse takes place (Ascending Node or Descending Node).
Geocentric Coordinates of the Sun and the Moon at Greatest Eclipse
    R.A. – Right Ascension   **S.D.** – Semi-Diameter (i.e. - radius)
    Dec. – Declination   **H.P.** – Horizontal Parallax
Eclipse Durations – Durations of Penumbral, Partial, and Total Eclipse.
Eclipse Contacts – Contact Times (Universal Time or UT1) of the Moon with the Penumbra and the Umbra
    P1, P4 – Start and End of the Penumbral Eclipse
    U1, U4 – Start and End of the Partial Eclipse
    U2, U3 – Start and End of the Total Eclipse

---

[9] Universal Time or UT1 is the modern replacement for Greenwich Mean Time

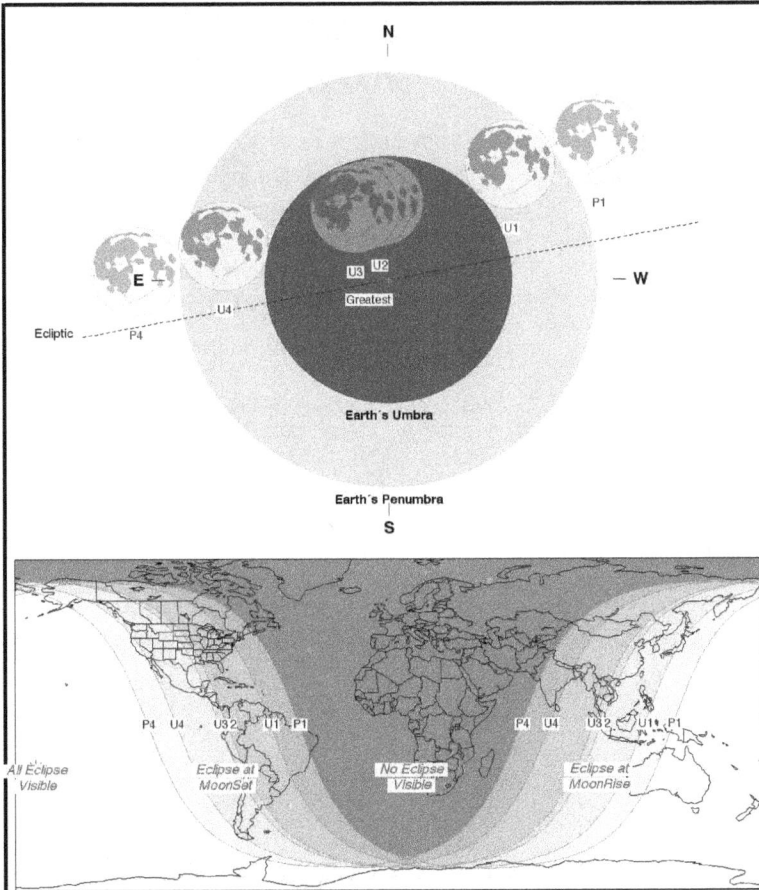

Total Lunar Eclipse
2021 May 26

Greatest Eclipse = 11:18:42.7 UT1
Penumbral Eclipse Magnitude = 1.9558
Umbral Eclipse Magnitude = 1.0112
Gamma = 0.4774
Axis = 0.4880°
Saros = 121 [55 of 82]
Decending Node

Geocentric Coordinates at Greatest Eclipse

Sun	Moon
R.A. = 04h14m03.6s	R.A. = 16h14m37.8s
Dec. = +21°12'25.4"	Dec. = -20°44'15.0"
S.D. = 00°15'47.3"	S.D. = 00°16'42.9"
H.P. = 00°00'08.7"	H.P. = 01°01'20.5"

Eclipse Durations	Eclipse Contacts
Penumbral = 05h02m50s	P1 = 08:47:21 UT1
Umbral = 03h08m07s	U1 = 09:44:43 UT1
Total = 00h15m52s	U2 = 11:10:56 UT1
	U3 = 11:26:48 UT1
	U4 = 12:52:50 UT1
Parameters	P4 = 13:50:11 UT1
Eph. = JPL DE430	
Rule = Herald-Sinnott	
ΔT = 70 s	

www.EclipseWise.com   ©2020 F. Espenak

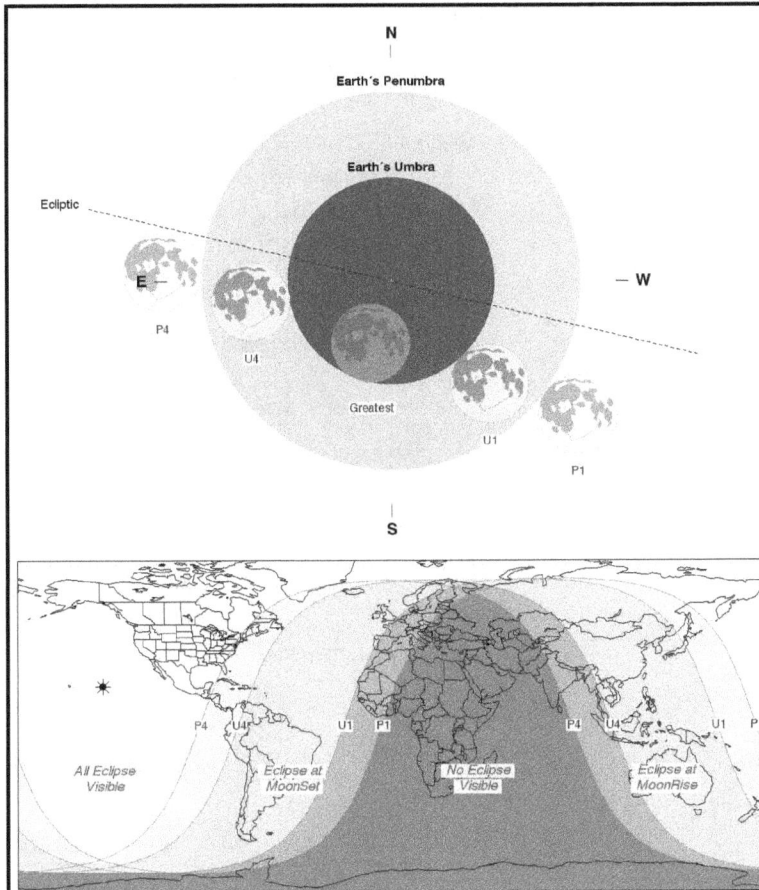

Partial Lunar Eclipse
2021 Nov 19

Greatest Eclipse = 09:02:55.5 UT1
Penumbral Eclipse Magnitude = 2.0738
Umbral Eclipse Magnitude = 0.9760
Gamma = -0.4552
Axis = 0.4104°
Saros = 126 [45 of 70]
Ascending Node

Geocentric Coordinates at Greatest Eclipse

Sun	Moon
R.A. = 15h39m50.9s	R.A. = 03h40m24.8s
Dec. = -19°32'33.1"	Dec. = +19°09'15.5"
S.D. = 00°16'11.0"	S.D. = 00°14'44.5"
H.P. = 00°00'08.9"	H.P. = 00°54'06.1"

Eclipse Durations	Eclipse Contacts
Penumbral = 06h02m21s	P1 = 06:01:49 UT1
Umbral = 03h29m09s	U1 = 07:18:26 UT1
	U4 = 10:47:36 UT1
	P4 = 12:04:11 UT1
Parameters	
Eph. = JPL DE430	
Rule = Herald-Sinnott	
ΔT = 70 s	

www.EclipseWise.com   ©2020 F. Espenak

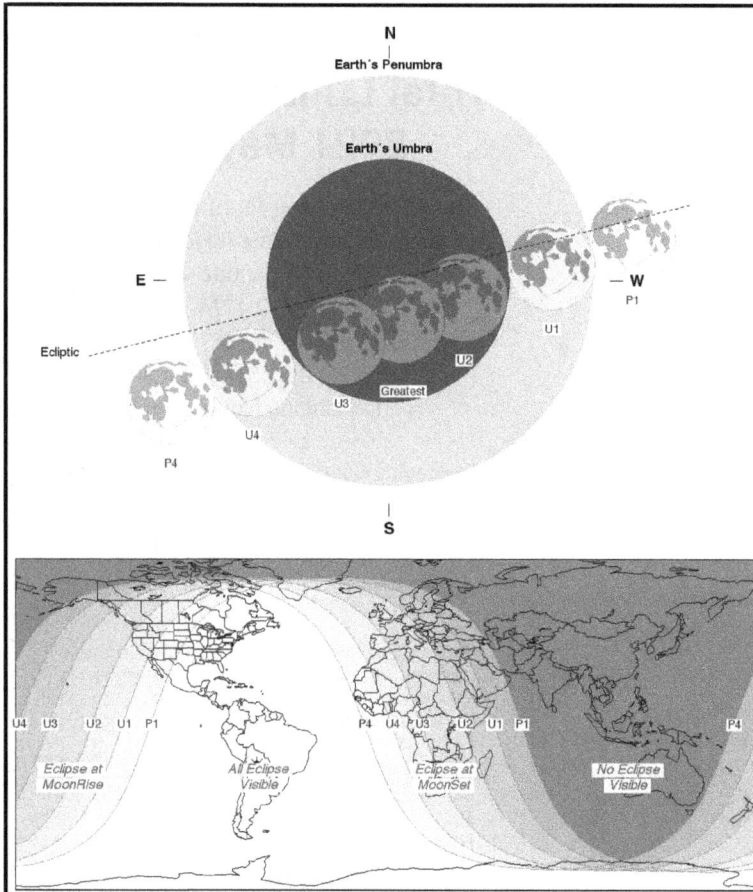

Total Lunar Eclipse
2022 May 16

Greatest Eclipse = 04:11:31.2 UT1
Penumbral Eclipse Magnitude = 2.3743
Umbral Eclipse Magnitude = 1.4154
Gamma = -0.2532
Axis = 0.2555°
Saros = 131 [34 of 72]
Decending Node

Geocentric Coordinates at Greatest Eclipse

Sun	Moon
R.A. = 03h31m49.5s	R.A. = 15h31m27.8s
Dec. = +19°05'13.4"	Dec. = -19°19'40.4"
S.D. = 00°15'49.2"	S.D. = 00°16'29.9"
H.P. = 00°00'08.7"	H.P. = 01°00'33.1"

Eclipse Durations	Eclipse Contacts
Penumbral = 05h19m28s	P1 = 01:31:44 UT1
Umbral = 03h27m57s	U1 = 02:27:31 UT1
Total = 01h25m32s	U2 = 03:28:40 UT1
	U3 = 04:54:11 UT1
	U4 = 05:55:27 UT1
Parameters	P4 = 06:51:11 UT1
Eph. = JPL DE430	
Rule = Herald-Sinnott	
ΔT = 70 s	

www.EclipseWise.com   ©2020 F. Espenak

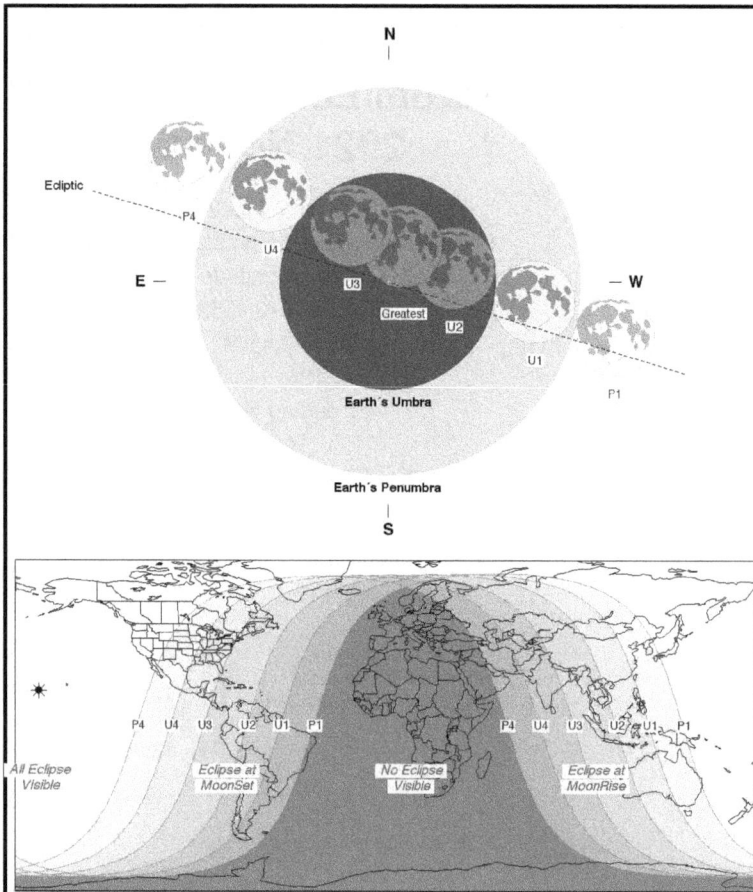

Total Lunar Eclipse
2022 Nov 08

Greatest Eclipse = 10:59:11.3 UT1
Penumbral Eclipse Magnitude = 2.4161
Umbral Eclipse Magnitude = 1.3607
Gamma = 0.2570
Axis = 0.2404°
Saros = 136 [20 of 72]
Ascending Node

Geocentric Coordinates at Greatest Eclipse

Sun	Moon
R.A. = 14h54m11.2s	R.A. = 02h53m48.1s
Dec. = -16°37'47.0"	Dec. = +16°51'06.7"
S.D. = 00°16'08.5"	S.D. = 00°15'17.7"
H.P. = 00°00'08.9"	H.P. = 00°56'07.8"

Eclipse Durations	Eclipse Contacts
Penumbral = 05h54m41s	P1 = 08:01:52 UT1
Umbral = 03h40m35s	U1 = 09:08:49 UT1
Total = 01h25m40s	U2 = 10:16:12 UT1
	U3 = 11:41:52 UT1
	U4 = 12:49:24 UT1
Parameters	P4 = 13:56:32 UT1
Eph. = JPL DE430	
Rule = Herald-Sinnott	
ΔT = 71 s	

www.EclipseWise.com   ©2020 F. Espenak

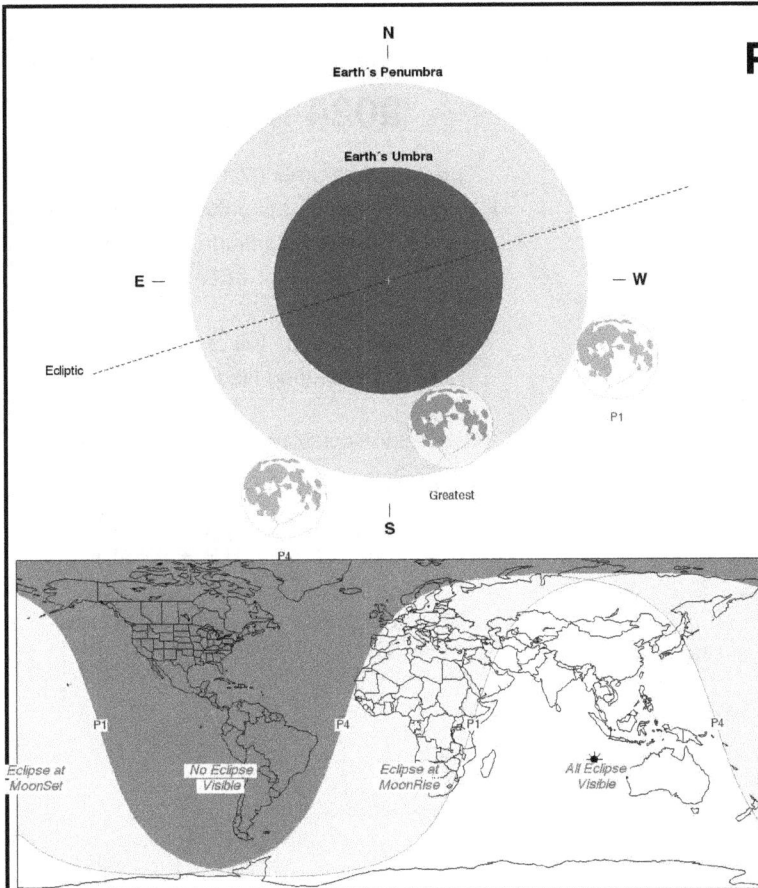

Penumbral Lunar Eclipse
2023 May 05

Greatest Eclipse = 17:22:53.5 UT1
Penumbral Eclipse Magnitude = 0.9655
Umbral Eclipse Magnitude = -0.0438
Gamma = -1.0350
Axis = 0.9946°
Saros = 141 [24 of 72]
Decending Node

Geocentric Coordinates at Greatest Eclipse

Sun	Moon
R.A. = 02h49m59.7s	R.A. = 14h48m23.5s
Dec. = +16°19'27.9"	Dec. = -17°14'31.7"
S.D. = 00°15'51.6"	S.D. = 00°15'42.8"
H.P. = 00°00'08.7"	H.P. = 00°57'40.1"

Eclipse Durations	Eclipse Contacts
Penumbral = 04h18m16s	P1 = 15:13:39 UT1
	P4 = 19:31:54 UT1

Parameters
Eph. = JPL DE430
Rule = Herald-Sinnott
ΔT = 71 s

www.EclipseWise.com   ©2020 F. Espenak

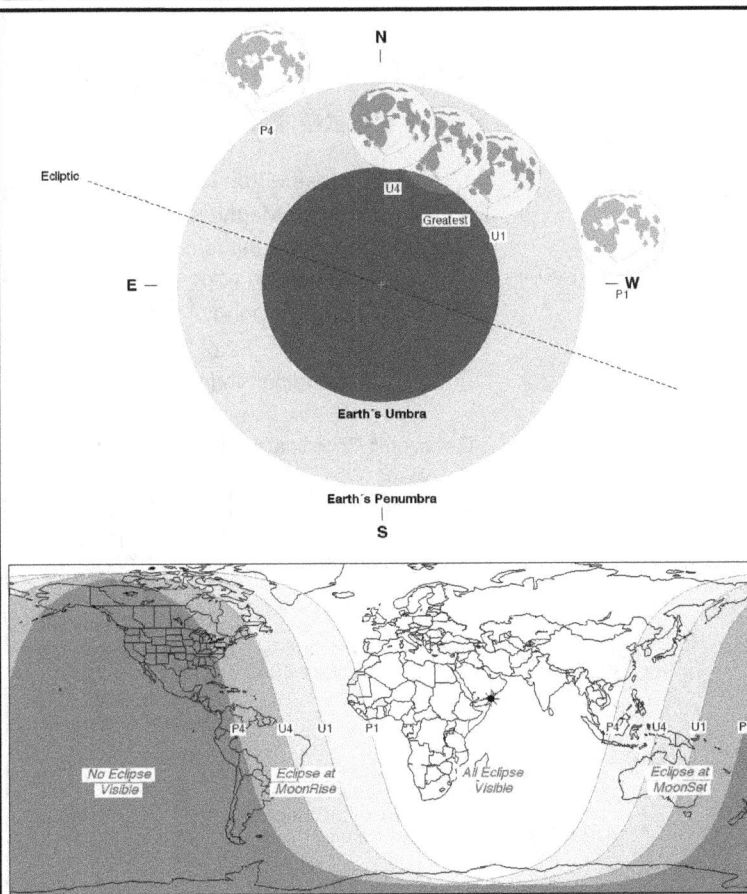

Partial Lunar Eclipse
2023 Oct 28

Greatest Eclipse = 20:14:05.9 UT1
Penumbral Eclipse Magnitude = 1.1200
Umbral Eclipse Magnitude = 0.1239
Gamma = 0.9472
Axis = 0.9362°
Saros = 146 [11 of 72]
Ascending Node

Geocentric Coordinates at Greatest Eclipse

Sun	Moon
R.A. = 14h11m25.9s	R.A. = 02h09m47.6s
Dec. = -13°14'10.5"	Dec. = +14°05'01.6"
S.D. = 00°16'05.9"	S.D. = 00°16'09.7"
H.P. = 00°00'08.9"	H.P. = 00°59'18.9"

Eclipse Durations	Eclipse Contacts
Penumbral = 04h25m16s	P1 = 18:01:17 UT1
Umbral = 01h18m09s	U1 = 19:34:41 UT1
	U4 = 20:52:50 UT1
	P4 = 22:26:33 UT1

Parameters
Eph. = JPL DE430
Rule = Herald-Sinnott
ΔT = 71 s

www.EclipseWise.com   ©2020 F. Espenak

Penumbral Lunar Eclipse
2024 Mar 25

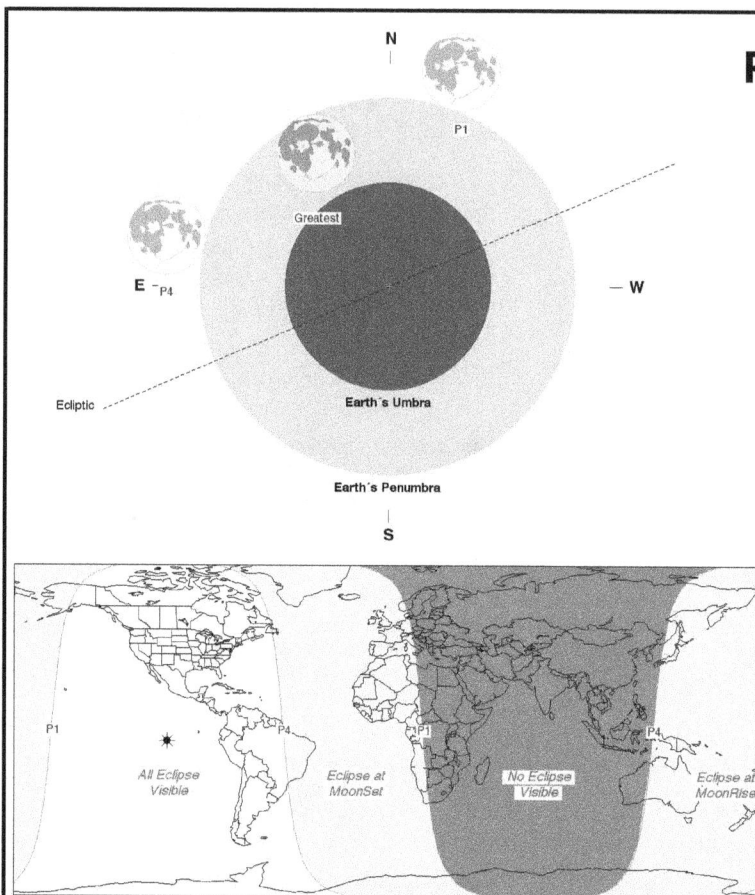

Greatest Eclipse = 07:12:48.8 UT1
Penumbral Eclipse Magnitude = 0.9577
Umbral Eclipse Magnitude = -0.1304
Gamma = 1.0610
Axis = 0.9564°
Saros = 113 [64 of 71]
Decending Node

Geocentric Coordinates at Greatest Eclipse

Sun	Moon
R.A. = 00h18m49.9s	R.A. = 12h20m41.3s
Dec. = +02°02'16.6"	Dec. = -01°12'05.6"
S.D. = 00°16'02.2"	S.D. = 00°14'44.3"
H.P. = 00°00'08.8"	H.P. = 00°54'05.4"

Eclipse Durations	Eclipse Contacts
Penumbral = 04h39m52s	P1 = 04:53:07 UT1
	P4 = 09:32:59 UT1

Parameters
Eph. = JPL DE430
Rule = Herald-Sinnott
ΔT = 71 s

www.EclipseWise.com   ©2020 F. Espenak

Partial Lunar Eclipse
2024 Sep 18

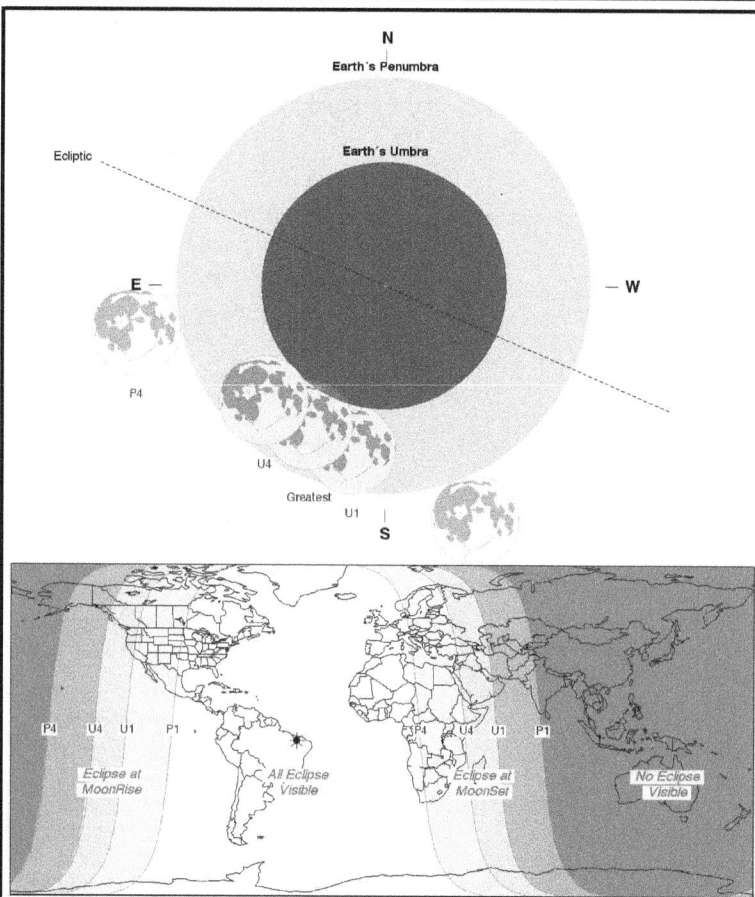

Greatest Eclipse = 02:44:14.1 UT1
Penumbral Eclipse Magnitude = 1.0392
Umbral Eclipse Magnitude = 0.0869
Gamma = -0.9792
Axis = 1.0009°
Saros = 118 [52 of 73]
Ascending Node

Geocentric Coordinates at Greatest Eclipse

Sun	Moon
R.A. = 11h44m09.7s	R.A. = 23h46m06.1s
Dec. = +01°42'52.9"	Dec. = -02°35'26.7"
S.D. = 00°15'55.1"	S.D. = 00°16'42.8"
H.P. = 00°00'08.8"	H.P. = 01°01'20.4"

Eclipse Durations	Eclipse Contacts
Penumbral = 04h06m56s	P1 = 00:40:58 UT1
Umbral = 01h03m40s	U1 = 02:12:42 UT1
	U4 = 03:16:22 UT1
	P4 = 04:47:54 UT1

Parameters
Eph. = JPL DE430
Rule = Herald-Sinnott
ΔT = 71 s

www.EclipseWise.com   ©2020 F. Espenak

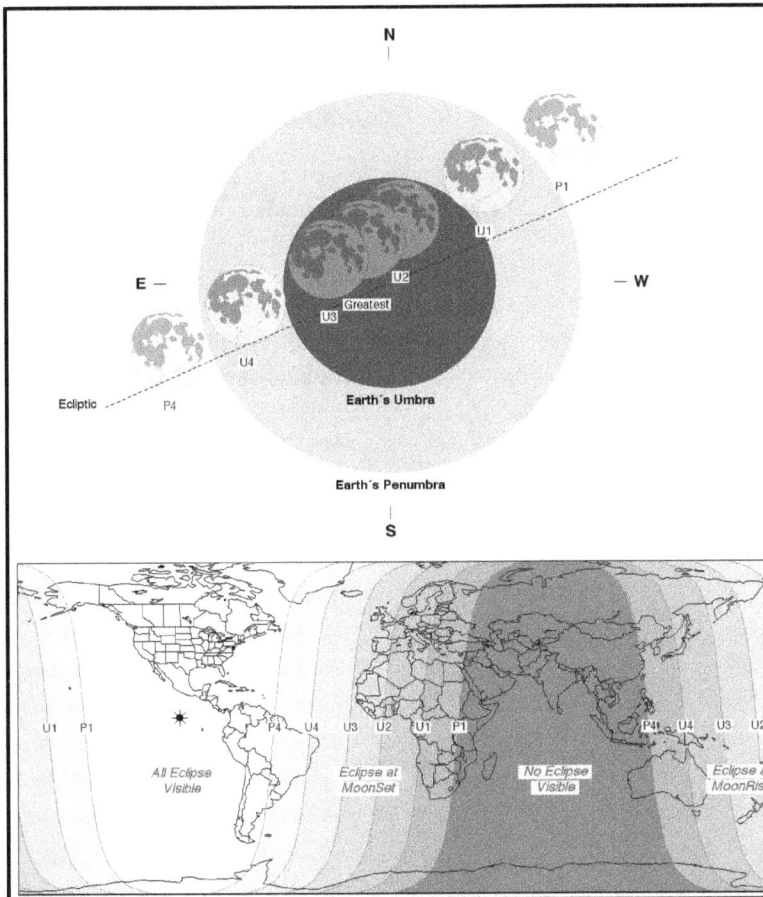

Total Lunar Eclipse
2025 Mar 14

Greatest Eclipse = 06:58:44.5 UT1
Penumbral Eclipse Magnitude = 2.2615
Umbral Eclipse Magnitude = 1.1804
Gamma = 0.3485
Axis = 0.3171°
Saros = 123 [53 of 72]
Decending Node

Geocentric Coordinates at Greatest Eclipse

Sun	Moon
R.A. = 23h37m46.0s	R.A. = 11h38m23.0s
Dec. = -02°24'16.8"	Dec. = +02°40'54.6"
S.D. = 00°16'05.2"	S.D. = 00°14'52.8"
H.P. = 00°00'08.8"	H.P. = 00°54'36.8"

Eclipse Durations
Penumbral = 06h03m22s
Umbral = 03h38m56s
Total = 01h06m04s

Eclipse Contacts
P1 = 03:57:09 UT1
U1 = 05:09:23 UT1
U2 = 06:25:58 UT1
U3 = 07:32:01 UT1
U4 = 08:48:18 UT1
P4 = 10:00:32 UT1

Parameters
Eph. = JPL DE430
Rule = Herald-Sinnott
ΔT = 72 s

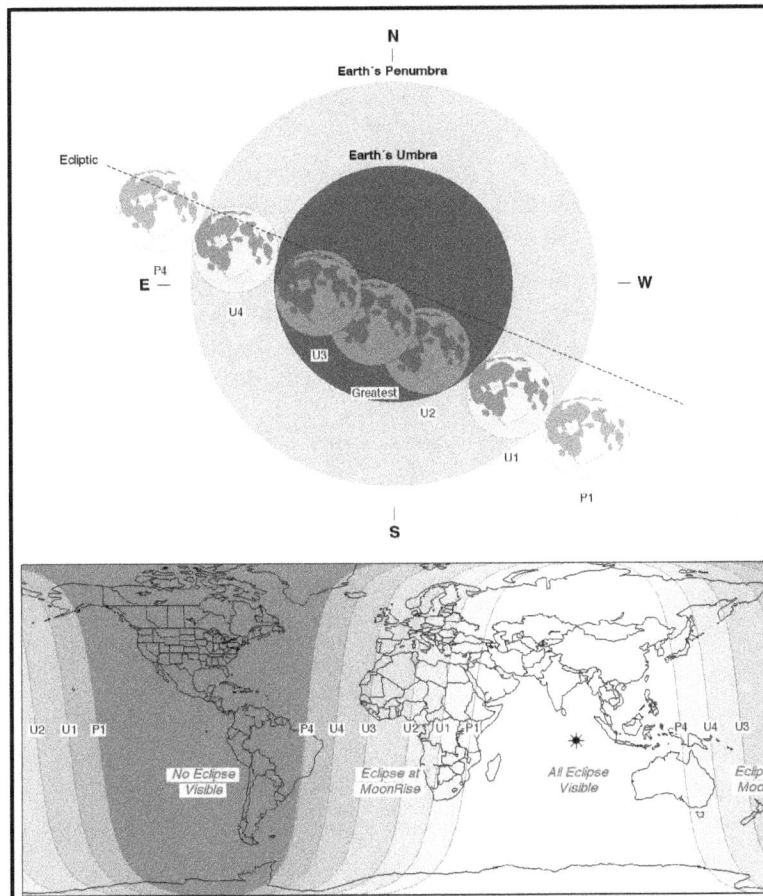

Total Lunar Eclipse
2025 Sep 07

Greatest Eclipse = 18:11:46.1 UT1
Penumbral Eclipse Magnitude = 2.3459
Umbral Eclipse Magnitude = 1.3638
Gamma = -0.2752
Axis = 0.2721°
Saros = 128 [41 of 71]
Ascending Node

Geocentric Coordinates at Greatest Eclipse

Sun	Moon
R.A. = 11h06m09.1s	R.A. = 23h06m40.4s
Dec. = +05°45'47.6"	Dec. = -06°00'08.9"
S.D. = 00°15'52.4"	S.D. = 00°16'09.8"
H.P. = 00°00'08.7"	H.P. = 00°59'19.1"

Eclipse Durations
Penumbral = 05h27m22s
Umbral = 03h30m02s
Total = 01h22m41s

Eclipse Contacts
P1 = 15:28:06 UT1
U1 = 16:26:51 UT1
U2 = 17:30:37 UT1
U3 = 18:53:18 UT1
U4 = 19:56:53 UT1
P4 = 20:55:28 UT1

Parameters
Eph. = JPL DE430
Rule = Herald-Sinnott
ΔT = 72 s

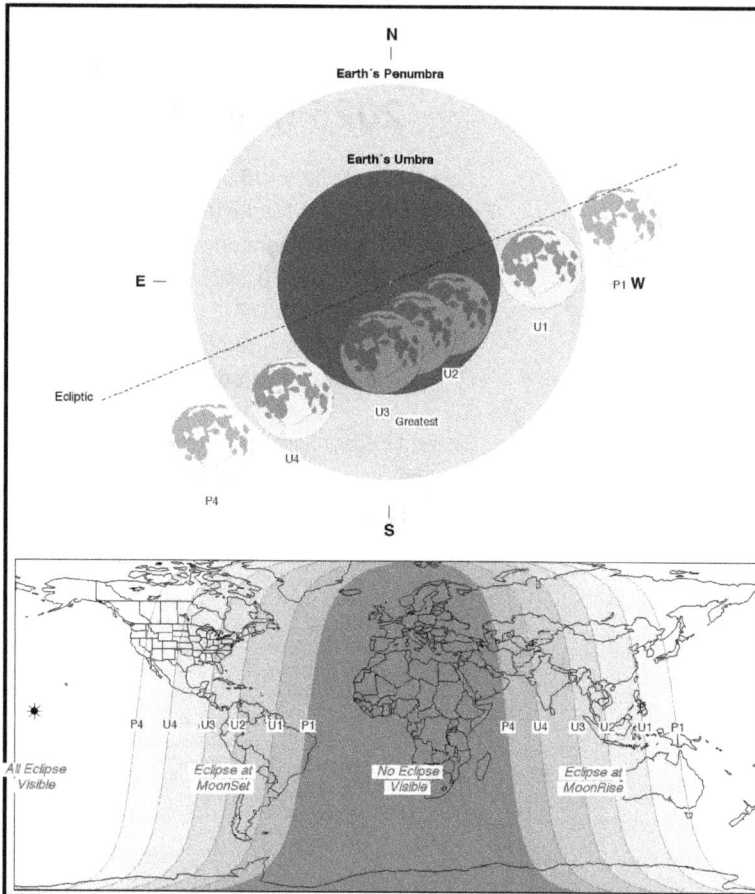

Total Lunar Eclipse
2026 Mar 03

Greatest Eclipse = 11:33:40.0 UT1
Penumbral Eclipse Magnitude = 2.1858
Umbral Eclipse Magnitude = 1.1526
Gamma = -0.3765
Axis = 0.3596°
Saros = 133 [27 of 71]
Decending Node

Geocentric Coordinates at Greatest Eclipse

Sun	Moon
R.A. = 22h56m56.0s	R.A. = 10h56m15.0s
Dec. = -06°43'06.4"	Dec. = +06°24'05.3"
S.D. = 00°16'08.0"	S.D. = 00°15'37.0"
H.P. = 00°00'08.9"	H.P. = 00°57'18.7"

Eclipse Durations	Eclipse Contacts
Penumbral = 05h39m21s	P1 = 08:43:58 UT1
Umbral = 03h27m49s	U1 = 09:49:37 UT1
Total = 00h58m58s	U2 = 11:03:54 UT1
	U3 = 12:02:53 UT1
	U4 = 13:17:26 UT1
Parameters	P4 = 14:23:19 UT1
Eph. = JPL DE430	
Rule = Herald-Sinnott	
ΔT = 72 s	

www.EclipseWise.com   ©2020 F. Espenak

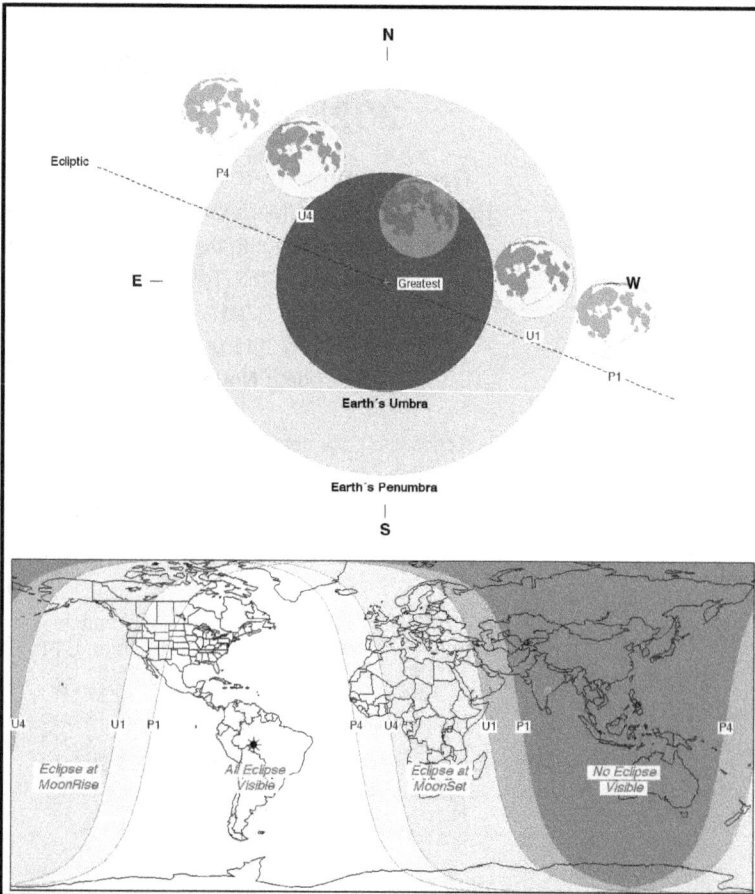

Partial Lunar Eclipse
2026 Aug 28

Greatest Eclipse = 04:12:52.0 UT1
Penumbral Eclipse Magnitude = 1.9664
Umbral Eclipse Magnitude = 0.9319
Gamma = 0.4964
Axis = 0.4647°
Saros = 138 [29 of 82]
Ascending Node

Geocentric Coordinates at Greatest Eclipse

Sun	Moon
R.A. = 10h26m57.9s	R.A. = 22h26m06.3s
Dec. = +09°42'52.7"	Dec. = -09°18'03.6"
S.D. = 00°15'50.0"	S.D. = 00°15'18.2"
H.P. = 00°00'08.7"	H.P. = 00°56'09.9"

Eclipse Durations	Eclipse Contacts
Penumbral = 05h38m31s	P1 = 01:23:29 UT1
Umbral = 03h18m48s	U1 = 02:33:21 UT1
	U4 = 05:52:09 UT1
	P4 = 07:02:00 UT1
Parameters	
Eph. = JPL DE430	
Rule = Herald-Sinnott	
ΔT = 72 s	

www.EclipseWise.com   ©2020 F. Espenak

Penumbral Lunar Eclipse
2027 Feb 20

Greatest Eclipse = 23:12:52.7 UT1

Penumbral Eclipse Magnitude = 0.9286

Umbral Eclipse Magnitude = -0.0549

Gamma = -1.0480

Axis = 1.0542°

Saros = 143 [18 of 72]

Decending Node

Geocentric Coordinates at Greatest Eclipse

Sun	Moon
R.A. = 22h16m18.3s	R.A. = 10h14m23.7s
Dec. = -10°43'53.9"	Dec. = +09°47'16.6"
S.D. = 00°16'10.5"	S.D. = 00°16'26.8"
H.P. = 00°00'08.9"	H.P. = 01°00'21.6"

Eclipse Durations	Eclipse Contacts
Penumbral = 04h01m41s	P1 = 21:11:49 UT1
	P4 = 01:13:30 UT1

Parameters
Eph. = JPL DE430

Rule = Herald-Sinnott

ΔT = 73 s

www.EclipseWise.com   ©2020 F. Espenak

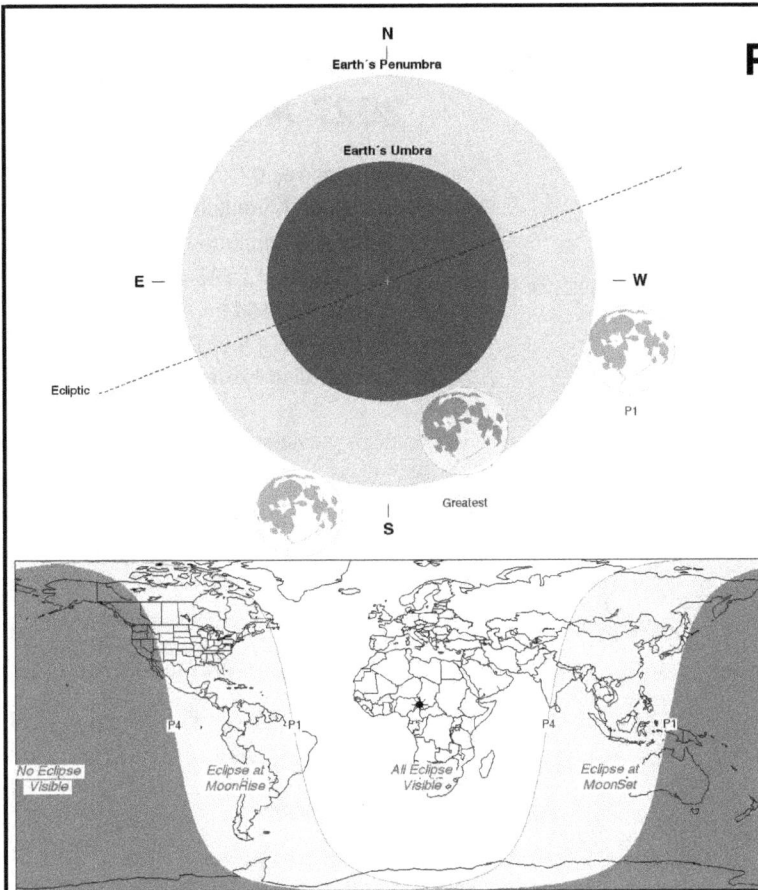

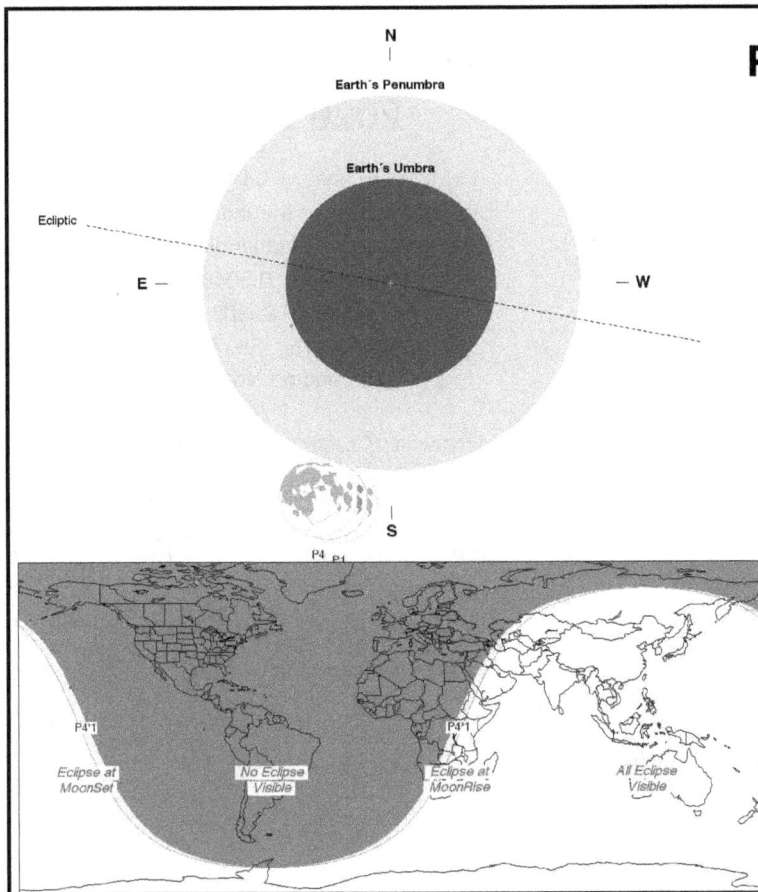

Penumbral Lunar Eclipse
2027 Jul 18

Greatest Eclipse = 16:02:57.8 UT1

Penumbral Eclipse Magnitude = 0.0032

Umbral Eclipse Magnitude = -1.0662

Gamma = -1.5759

Axis = 1.4184°

Saros = 110 [72 of 72]

Ascending Node

Geocentric Coordinates at Greatest Eclipse

Sun	Moon
R.A. = 07h51m14.4s	R.A. = 19h52m57.2s
Dec. = +20°58'43.6"	Dec. = -22°20'25.3"
S.D. = 00°15'44.3"	S.D. = 00°14'43.0"
H.P. = 00°00'08.7"	H.P. = 00°54'00.6"

Eclipse Durations	Eclipse Contacts
Penumbral = 00h17m39s	P1 = 15:54:18 UT1
	P4 = 16:11:57 UT1

Parameters
Eph. = JPL DE430

Rule = Herald-Sinnott

ΔT = 73 s

www.EclipseWise.com   ©2020 F. Espenak

Penumbral Lunar Eclipse
2027 Aug 17

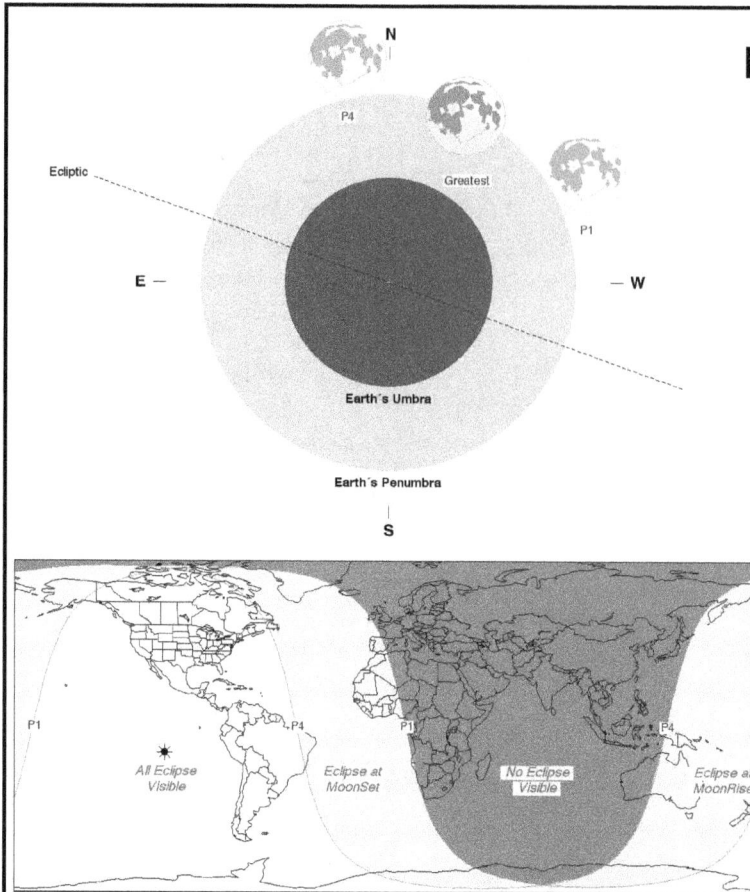

Greatest Eclipse = 07:13:45.1 UT1
Penumbral Eclipse Magnitude = 0.5476
Umbral Eclipse Magnitude = -0.5234
Gamma = 1.2797
Axis = 1.1544°
Saros = 148 [4 of 70]
Ascending Node

Geocentric Coordinates at Greatest Eclipse

Sun	Moon
R.A. = 09h45m58.6s	R.A. = 21h43m58.8s
Dec. = +13°27'30.2"	Dec. = -12°24'40.9"
S.D. = 00°15'47.8"	S.D. = 00°14'44.9"
H.P. = 00°00'08.7"	H.P. = 00°54'07.8"

Eclipse Durations	Eclipse Contacts
Penumbral = 03h39m21s	P1 = 05:23:52 UT1
	P4 = 09:03:14 UT1

Parameters
Eph. = JPL DE430
Rule = Herald-Sinnott
ΔT = 73 s

www.EclipseWise.com   ©2020 F. Espenak

Partial Lunar Eclipse
2028 Jan 12

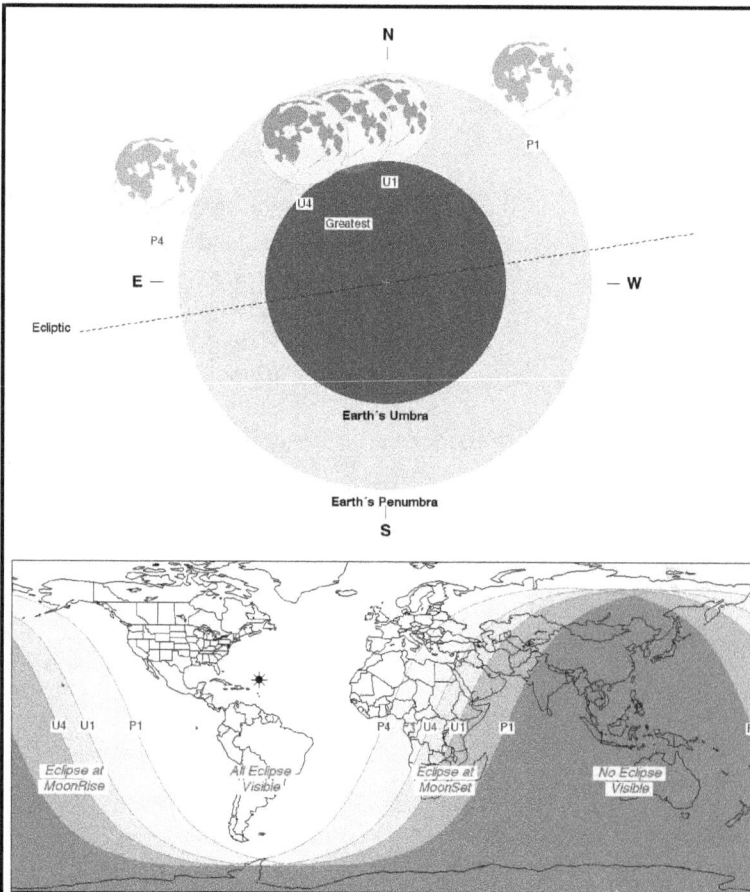

Greatest Eclipse = 04:13:00.5 UT1
Penumbral Eclipse Magnitude = 1.0485
Umbral Eclipse Magnitude = 0.0679
Gamma = 0.9818
Axis = 0.9958°
Saros = 115 [58 of 72]
Decending Node

Geocentric Coordinates at Greatest Eclipse

Sun	Moon
R.A. = 19h32m47.8s	R.A. = 07h33m53.0s
Dec. = -21°43'29.4"	Dec. = +22°41'18.2"
S.D. = 00°16'15.8"	S.D. = 00°16'35.1"
H.P. = 00°00'08.9"	H.P. = 01°00'52.0"

Eclipse Durations	Eclipse Contacts
Penumbral = 04h11m26s	P1 = 02:07:24 UT1
Umbral = 00h56m53s	U1 = 03:44:46 UT1
	U4 = 04:41:39 UT1
	P4 = 06:18:50 UT1

Parameters
Eph. = JPL DE430
Rule = Herald-Sinnott
ΔT = 73 s

www.EclipseWise.com   ©2020 F. Espenak

Partial Lunar Eclipse
2028 Jul 06

Greatest Eclipse = 18:19:44.0 UT1
Penumbral Eclipse Magnitude = 1.4282
Umbral Eclipse Magnitude = 0.3908
Gamma = -0.7904
Axis = 0.7331°
Saros = 120   [58 of 83]
Ascending Node

Geocentric Coordinates at Greatest Eclipse

Sun	Moon
R.A. = 07h05m56.7s	R.A. = 19h06m37.0s
Dec. = +22°34'16.5"	Dec. = -23°17'16.4"
S.D. = 00°15'43.9"	S.D. = 00°15'09.9"
H.P. = 00°00'08.6"	H.P. = 00°55'39.4"

Eclipse Durations	Eclipse Contacts
Penumbral = 05h11m29s	P1 = 15:44:03 UT1
Umbral = 02h22m13s	U1 = 17:08:40 UT1
	U4 = 19:30:53 UT1
	P4 = 20:55:33 UT1

Parameters
Eph. = JPL DE430
Rule = Herald-Sinnott
ΔT =   73 s

www.EclipseWise.com   ©2020 F. Espenak

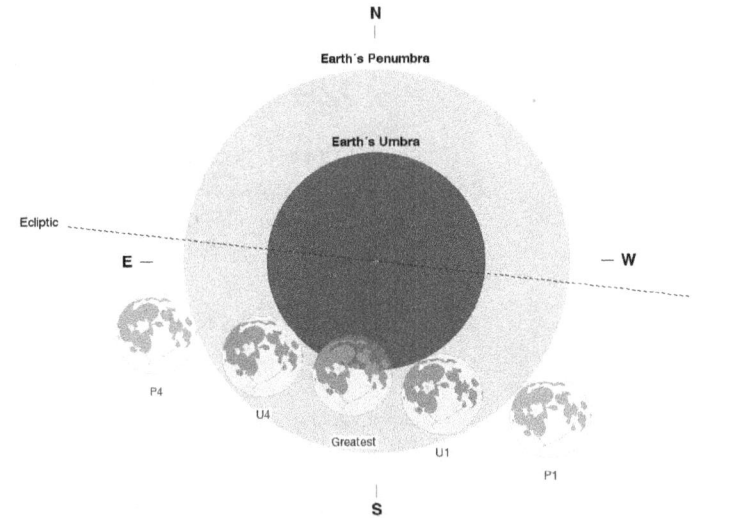

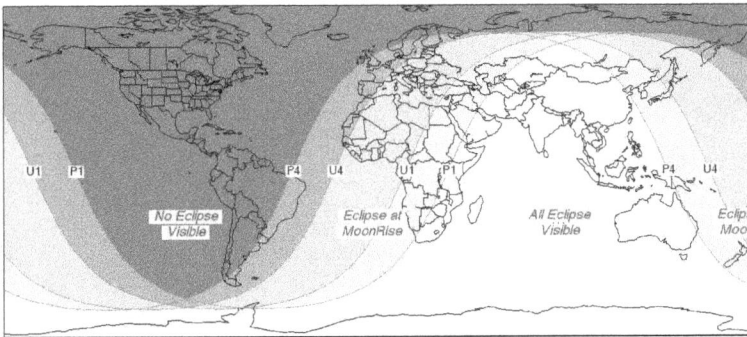

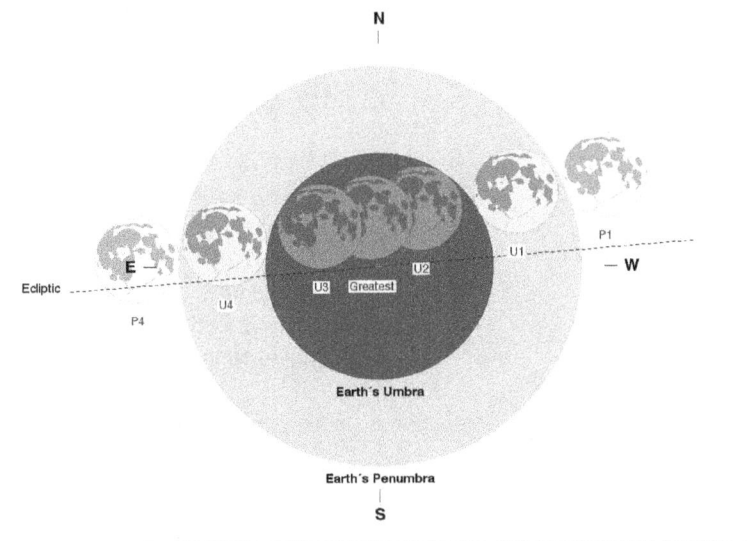

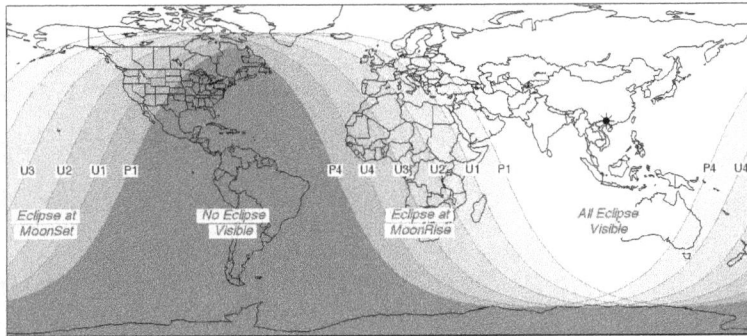

Total Lunar Eclipse
2028 Dec 31

Greatest Eclipse = 16:52:01.7 UT1
Penumbral Eclipse Magnitude = 2.2758
Umbral Eclipse Magnitude = 1.2478
Gamma = 0.3258
Axis = 0.3153°
Saros = 125   [49 of 72]
Decending Node

Geocentric Coordinates at Greatest Eclipse

Sun	Moon
R.A. = 18h45m53.7s	R.A. = 06h46m08.4s
Dec. = -23°01'00.5"	Dec. = +23°19'37.5"
S.D. = 00°16'15.9"	S.D. = 00°15'49.4"
H.P. = 00°00'08.9"	H.P. = 00°58'04.3"

Eclipse Durations	Eclipse Contacts
Penumbral = 05h37m06s	P1 = 14:03:27 UT1
Umbral = 03h29m37s	U1 = 15:07:17 UT1
Total = 01h12m02s	U2 = 16:16:07 UT1
	U3 = 17:28:08 UT1
	U4 = 18:36:54 UT1
	P4 = 19:40:34 UT1

Parameters
Eph. = JPL DE430
Rule = Herald-Sinnott
ΔT =   73 s

www.EclipseWise.com   ©2020 F. Espenak

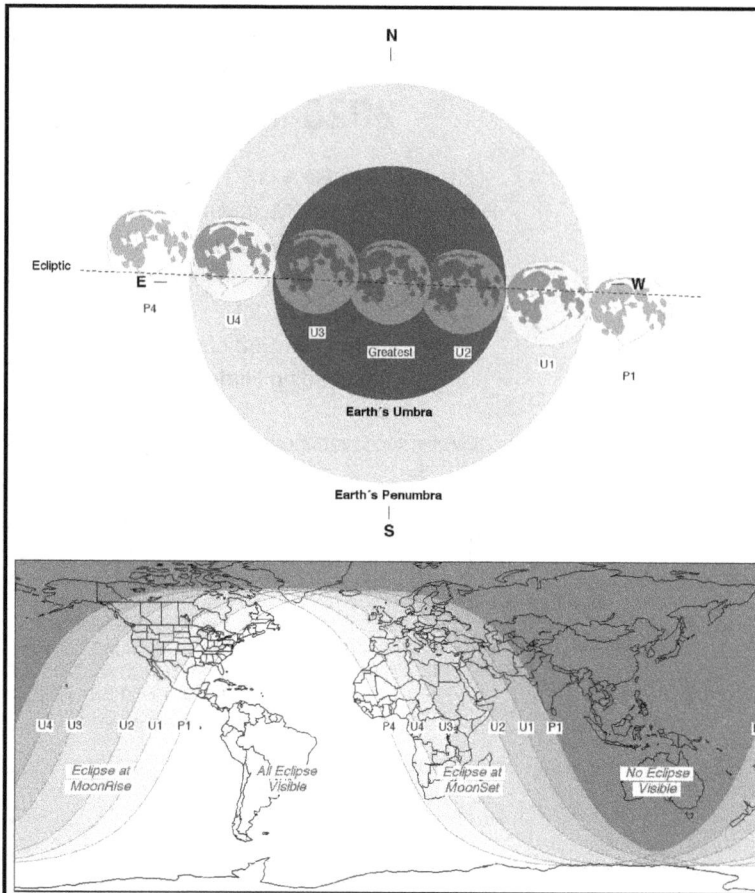

Total Lunar Eclipse
2029 Jun 26

Greatest Eclipse = 03:22:08.8 UT1
Penumbral Eclipse Magnitude = 2.8282
Umbral Eclipse Magnitude = 1.8452
Gamma = 0.0124
Axis = 0.0121°
Saros = 130 [35 of 71]
Ascending Node

Geocentric Coordinates at Greatest Eclipse

Sun	Moon
R.A. = 06h21m03.1s	R.A. = 18h21m02.6s
Dec. = +23°20'50.2"	Dec. = -23°20'06.9"
S.D. = 00°15'44.1"	S.D. = 00°16'00.4"
H.P. = 00°00'08.7"	H.P. = 00°58'44.7"

Eclipse Durations
Penumbral = 05h36m01s
Umbral = 03h40m20s
Total = 01h42m40s

Eclipse Contacts
P1 = 00:34:11 UT1
U1 = 01:31:58 UT1
U2 = 02:30:48 UT1
U3 = 04:13:28 UT1
U4 = 05:12:18 UT1
P4 = 06:10:12 UT1

Parameters
Eph. = JPL DE430
Rule = Herald-Sinnott
ΔT =   74 s

www.EclipseWise.com   ©2020 F. Espenak

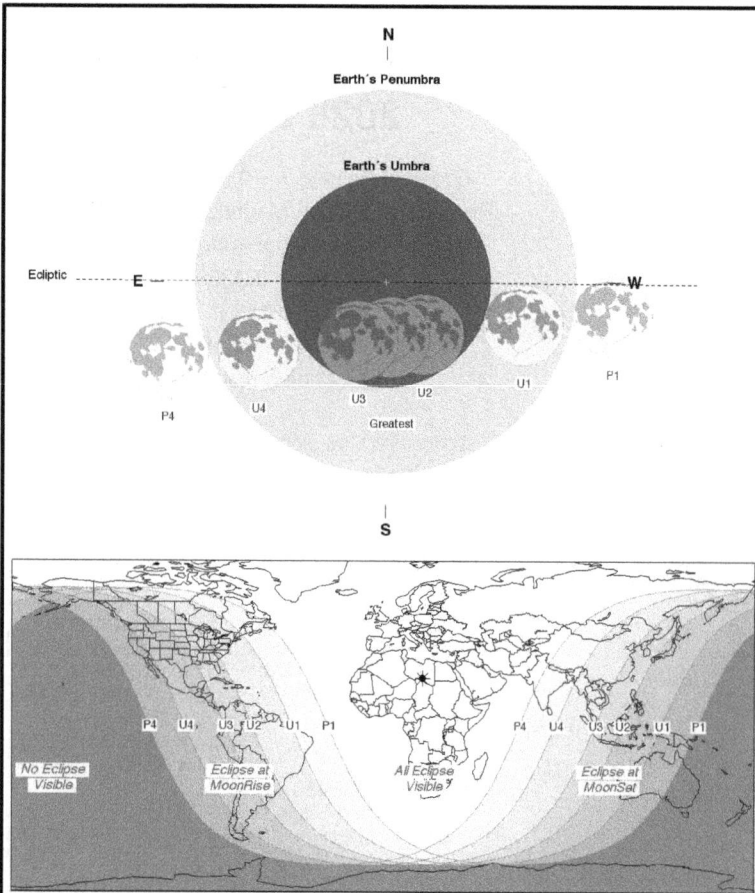

Total Lunar Eclipse
2029 Dec 20

Greatest Eclipse = 22:41:57.9 UT1
Penumbral Eclipse Magnitude = 2.2023
Umbral Eclipse Magnitude = 1.1190
Gamma = -0.3811
Axis = 0.3498°
Saros = 135 [24 of 71]
Decending Node

Geocentric Coordinates at Greatest Eclipse

Sun	Moon
R.A. = 17h57m07.6s	R.A. = 05h56m59.0s
Dec. = -23°26'00.2"	Dec. = +23°05'06.7"
S.D. = 00°16'15.5"	S.D. = 00°15'00.4"
H.P. = 00°00'08.9"	H.P. = 00°55'04.6"

Eclipse Durations
Penumbral = 05h58m54s
Umbral = 03h34m07s
Total = 00h54m23s

Eclipse Contacts
P1 = 19:42:28 UT1
U1 = 20:54:54 UT1
U2 = 22:14:44 UT1
U3 = 23:09:07 UT1
U4 = 00:29:01 UT1
P4 = 01:41:22 UT1

Parameters
Eph. = JPL DE430
Rule = Herald-Sinnott
ΔT =   74 s

www.EclipseWise.com   ©2020 F. Espenak

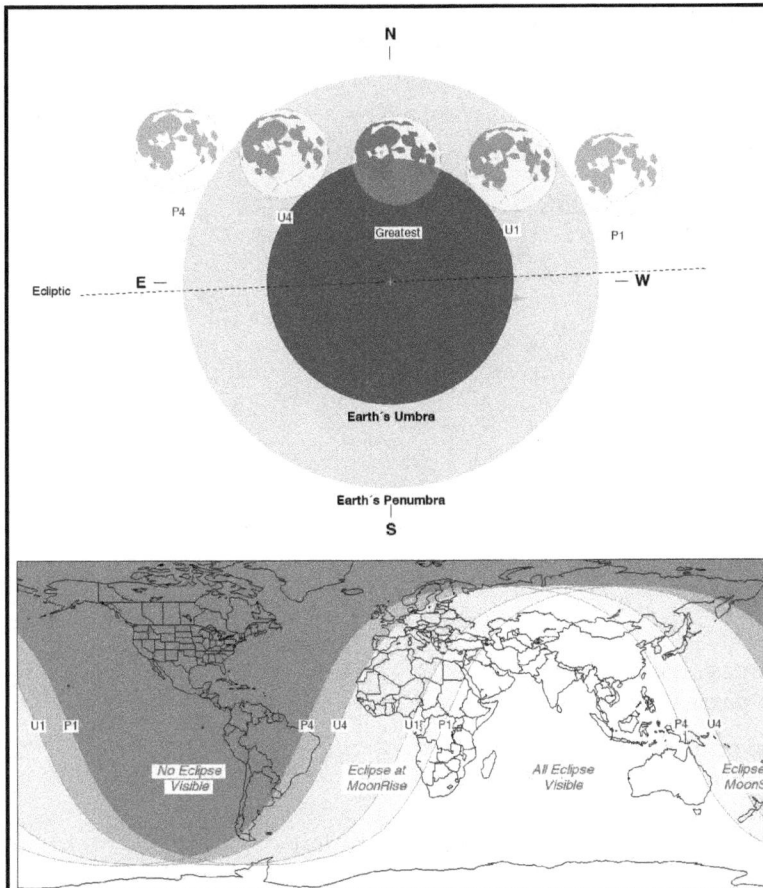

Partial Lunar Eclipse
2030 Jun 15

Greatest Eclipse = 18:33:19.4 UT1
Penumbral Eclipse Magnitude = 1.4495
Umbral Eclipse Magnitude = 0.5040
Gamma = 0.7535
Axis = 0.7674°
Saros = 140 [25 of 77]
Ascending Node

Geocentric Coordinates at Greatest Eclipse

Sun	Moon
R.A. = 05h36m57.6s	R.A. = 17h36m46.1s
Dec. = +23°19'44.0"	Dec. = -22°33'45.8"
S.D. = 00°15'44.7"	S.D. = 00°16'39.2"
H.P. = 00°00'08.7"	H.P. = 01°01'07.1"

Eclipse Durations	Eclipse Contacts
Penumbral = 04h39m02s	P1 = 16:13:47 UT1
Umbral = 02h25m02s	U1 = 17:20:46 UT1
	U4 = 19:45:47 UT1
	P4 = 20:52:49 UT1

Parameters
Eph. = JPL DE430
Rule = Herald-Sinnott
ΔT = 74 s

www.EclipseWise.com   ©2020 F. Espenak

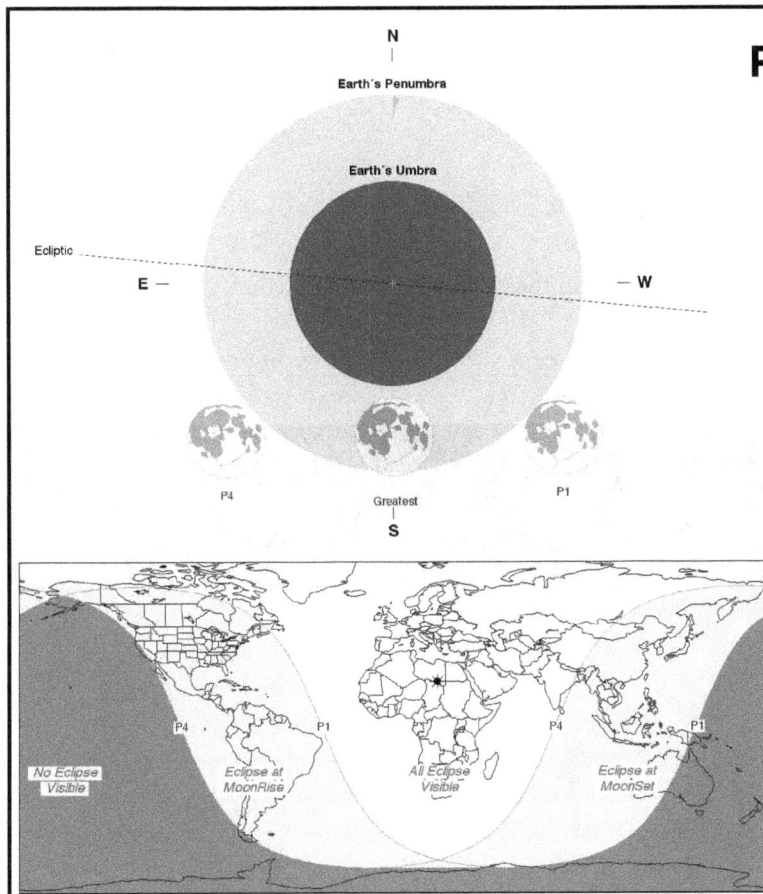

Penumbral Lunar Eclipse
2030 Dec 09

Greatest Eclipse = 22:27:36.7 UT1
Penumbral Eclipse Magnitude = 0.9430
Umbral Eclipse Magnitude = -0.1613
Gamma = -1.0732
Axis = 0.9652°
Saros = 145 [12 of 71]
Decending Node

Geocentric Coordinates at Greatest Eclipse

Sun	Moon
R.A. = 17h07m21.3s	R.A. = 05h07m19.1s
Dec. = -22°52'57.8"	Dec. = +21°55'03.1"
S.D. = 00°16'14.4"	S.D. = 00°14'42.3"
H.P. = 00°00'08.9"	H.P. = 00°53'58.2"

Eclipse Durations	Eclipse Contacts
Penumbral = 04h40m04s	P1 = 20:07:34 UT1
	P4 = 00:47:38 UT1

Parameters
Eph. = JPL DE430
Rule = Herald-Sinnott
ΔT = 74 s

www.EclipseWise.com   ©2020 F. Espenak

Key to Catalog of Lunar Eclipses

The following catalog lists all lunar eclipses during a 25-year period.

A brief description of each parameter in the catalog appears below.

Date — Gregorian date of Greatest Eclipse

Greatest Eclipse — Universal Time of Greatest Eclipse

Saros — Saros Series Number of the eclipse

Type — Lunar Eclipse Type

> N = Penumbral Lunar Eclipse
> P = Partial Lunar Eclipse
> T = Total Lunar Eclipse
>
> + = Central total eclipse (Moon's center passes north of shadow axis)
> – = Central total eclipse (Moon's center passes south of shadow axis)
>
> b = Saros series begins (first penumbral eclipse in a Saros series)
> e = Saros series ends (last penumbral eclipse in a Saros series)

Gamma — minimum distance from center of the Moon to axis of Earth's umbral shadow

P. Mag — Penumbral Magnitude; fraction of the Moon's diameter immersed in the penumbra

P. Mag — Umbral Magnitude; fraction of the Moon's diameter immersed in the umbra

Phase Durations

> **Pen** — elapsed time from contact P1 to P4 (minutes)
> **Par** — elapsed time from contact U1 to U4 (minutes)
> **Total** — elapsed time from contact U2 to U3 (minutes)

Lat & Long — latitude and longitude where the Moon appears in the zenith at Greatest Eclipse

Photo 2–8 Time sequence of the total lunar eclipse of 2014 April 15. ©2014 F. Espenak

Catalog of Lunar Eclipses: 2021 to 2045

Date	Greatest Eclipse	Saros	Type	Gamma	P. Mag	U.Mag	Pen	Par	Total	Lat.	Long.
2021 May 26	11:18:43	121	T	0.4774	1.9558	1.0112	302.8	188.1	15.9	20.7S	170.3W
2021 Nov 19	9:02:56	126	P	-0.4552	2.0738	0.9760	362.4	209.2	-	19.2N	139.2W
2022 May 16	4:11:31	131	T-	-0.2532	2.3743	1.4154	319.5	207.9	85.5	19.3S	63.9W
2022 Nov 08	10:59:11	136	T+	0.2570	2.4161	1.3607	354.7	220.6	85.7	16.9N	169.0W
2023 May 05	17:22:53	141	N	-1.0350	0.9655	-0.0438	258.3	-	-	17.2S	98.1E
2023 Oct 28	20:14:06	146	P	0.9472	1.1200	0.1239	265.3	78.1	-	14.1N	52.0E
2024 Mar 25	7:12:49	113	N	1.0610	0.9577	-0.1304	279.9	-	-	1.2S	106.3W
2024 Sep 18	2:44:14	118	P	-0.9792	1.0392	0.0869	246.9	63.7	-	2.6S	42.1W
2025 Mar 14	6:58:45	123	T	0.3485	2.2615	1.1804	363.4	218.9	66.1	2.7N	102.2W
2025 Sep 07	18:11:46	128	T	-0.2752	2.3459	1.3638	327.4	210.0	82.7	6.0S	86.7E
2026 Mar 03	11:33:40	133	T	-0.3765	2.1858	1.1526	339.4	207.8	59.0	6.4N	170.6W
2026 Aug 28	4:12:52	138	P	0.4964	1.9664	0.9319	338.5	198.8	-	9.3S	63.1W
2027 Feb 20	23:12:53	143	N	-1.0480	0.9286	-0.0549	241.7	-	-	9.8N	14.7E
2027 Jul 18	16:02:58	110	Ne	-1.5759	0.0032	-1.0662	17.7	-	-	22.3S	121.3E
2027 Aug 17	7:13:45	148	N	1.2797	0.5476	-0.5234	219.4	-	-	12.4S	107.9W
2028 Jan 12	4:13:00	115	P	0.9818	1.0485	0.0679	251.4	56.9	-	22.7N	61.0W
2028 Jul 06	18:19:44	120	P	-0.7904	1.4282	0.3908	311.5	142.2	-	23.3S	86.5E
2028 Dec 31	16:52:02	125	T	0.3258	2.2758	1.2478	337.1	209.6	72.0	23.3N	107.9E
2029 Jun 26	3:22:09	130	T+	0.0124	2.8282	1.8452	336.0	220.3	102.7	23.3S	49.8W
2029 Dec 20	22:41:58	135	T	-0.3811	2.2023	1.1190	358.9	214.1	54.4	23.1N	19.0E
2030 Jun 15	18:33:19	140	P	0.7535	1.4495	0.5040	279.0	145.0	-	22.6S	81.8E
2030 Dec 09	22:27:37	145	N	-1.0732	0.9430	-0.1613	280.1	-	-	21.9N	21.2E
2031 May 07	3:50:47	112	N	-1.0695	0.8827	-0.0892	238.1	-	-	17.8S	58.8W
2031 Jun 05	11:44:03	150	N	1.4732	0.1306	-0.8185	96.3	-	-	21.1S	176.3W
2031 Oct 30	7:45:29	117	N	1.1774	0.7173	-0.3193	232.5	-	-	14.8N	120.7W
2032 Apr 25	15:13:36	122	T	-0.3558	2.2204	1.1925	343.3	212.0	66.2	13.8S	131.0E
2032 Oct 18	19:02:25	127	T	0.4169	2.0841	1.1039	316.2	196.6	47.7	10.4N	70.5E
2033 Apr 14	19:12:36	132	T	0.3954	2.1722	1.0955	362.1	215.8	49.8	9.4S	72.0E
2033 Oct 08	10:55:07	137	T	-0.2889	2.3068	1.3508	313.4	203.1	79.4	5.8N	166.8W
2034 Apr 03	19:05:44	142	N	1.1144	0.8557	-0.2263	266.2	-	-	4.6S	74.7E
2034 Sep 28	2:46:21	147	P	-1.0110	0.9922	0.0155	249.4	27.8	-	1.0N	43.6W
2035 Feb 22	9:04:55	114	N	-1.0367	0.9663	-0.0523	256.4	-	-	9.2N	133.1W
2035 Aug 19	1:10:58	119	P	0.9434	1.1519	0.1049	290.6	77.2	-	12.0S	17.0W
2036 Feb 11	22:11:49	124	T	-0.3110	2.2762	1.3006	316.9	202.7	75.1	13.6N	30.5E
2036 Aug 07	2:51:15	129	T+	0.2004	2.5279	1.4556	373.1	232.2	96.1	16.1N	41.4W
2037 Jan 31	14:00:21	134	T	0.3619	2.1815	1.2086	312.9	198.2	64.3	17.5N	153.3E
2037 Jul 27	4:08:36	139	P	-0.5582	1.8596	0.8108	341.7	193.2	-	19.6S	60.4W
2038 Jan 21	3:48:34	144	N	1.0711	0.9009	-0.1127	246.5	-	-	20.9N	54.2W
2038 Jun 17	2:43:44	111	N	1.3083	0.4438	-0.5259	177.0	-	-	22.1S	40.5W
2038 Jul 16	11:34:38	149	N	-1.2838	0.5012	-0.4938	193.2	-	-	22.5S	172.0W
2038 Dec 11	17:43:42	116	N	-1.1449	0.8062	-0.2876	259.3	-	-	22.0N	92.6E
2039 Jun 06	18:53:07	121	P	0.5460	1.8288	0.8863	297.5	180.0	-	22.1S	76.5E
2039 Nov 30	16:55:09	126	P	-0.4721	2.0435	0.9443	361.0	206.8	-	21.3N	103.5E
2040 May 26	11:45:02	131	T-	-0.1872	2.4955	1.5365	322.2	211.5	92.9	21.5S	177.0W
2040 Nov 18	19:03:21	136	T+	0.2361	2.4543	1.3991	354.4	221.2	88.6	19.7N	70.4E
2041 May 16	0:41:42	141	P	-0.9747	1.0765	0.0663	270.5	59.4	-	20.0S	11.7W
2041 Nov 08	4:33:44	146	P	0.9212	1.1675	0.1714	268.7	91.1	-	17.5N	72.9W
2042 Apr 05	14:28:52	113	N	1.1080	0.8700	-0.2156	269.2	-	-	5.4S	143.9E
2042 Sep 29	10:44:27	118	N	-1.0262	0.9548	-0.0010	239.2	-	-	1.6N	163.0W
2043 Mar 25	14:30:43	123	T	0.3849	2.1920	1.1161	360.0	215.3	54.1	1.6S	144.0E
2043 Sep 19	1:50:29	128	T	-0.3316	2.2452	1.2575	326.4	206.7	72.3	1.9S	29.0W
2044 Mar 13	19:37:11	133	T	-0.3496	2.2322	1.2050	339.1	209.7	67.0	2.1N	67.8E
2044 Sep 07	11:19:23	138	T	0.4318	2.0879	1.0476	344.8	206.9	34.8	5.4S	170.6W
2045 Mar 03	7:42:03	143	N	-1.0274	0.9643	-0.0148	244.6	-	-	5.7N	113.0W
2045 Aug 27	13:53:26	148	N	1.2061	0.6845	-0.3899	242.4	-	-	8.8S	151.5E

Section 3: Phases of the Moon: 2021 to 2030

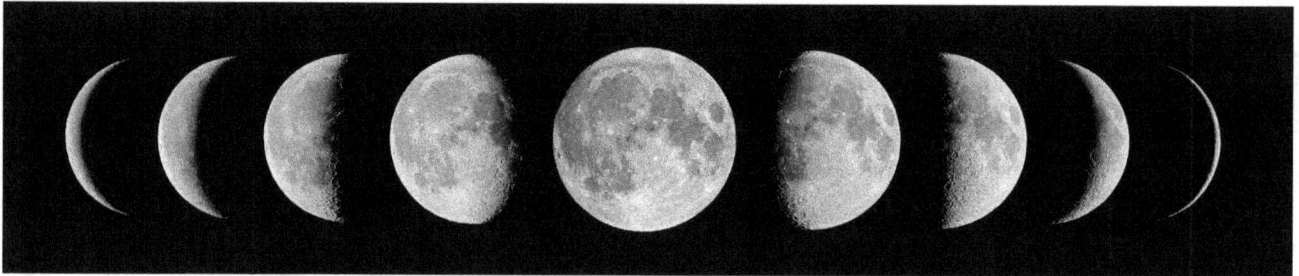

Photo 3–1 Various phases of the Moon over one lunar month. ©2010 F. Espenak

Phases of the Moon

Year	New Moon	First Quarter	Full Moon	Last Quarter
2021				Jan 06 09:37
	Jan 13 05:00	Jan 20 21:02	Jan 28 19:16	Feb 04 17:37
	Feb 11 19:06	Feb 19 18:47	Feb 27 08:17	Mar 06 01:30
	Mar 13 10:21	Mar 21 14:40	Mar 28 18:48	Apr 04 10:02
	Apr 12 02:31	Apr 20 06:59	Apr 27 03:31	May 03 19:50
	May 11 19:00	May 19 19:13	May 26 11:14 t	Jun 02 07:24
	Jun 10 10:53 A	Jun 18 03:54	Jun 24 18:40	Jul 01 21:11
	Jul 10 01:17	Jul 17 10:11	Jul 24 02:37	Jul 31 13:16
	Aug 08 13:50	Aug 15 15:20	Aug 22 12:02	Aug 30 07:13
	Sep 07 00:52	Sep 13 20:39	Sep 20 23:55	Sep 29 01:57
	Oct 06 11:05	Oct 13 03:25	Oct 20 14:57	Oct 28 20:05
	Nov 04 21:15	Nov 11 12:46	Nov 19 08:58 p	Nov 27 12:28
	Dec 04 07:43 T	Dec 11 01:36	Dec 19 04:36	Dec 27 02:24

Year	New Moon	First Quarter	Full Moon	Last Quarter
2022	Jan 02 18:33	Jan 09 18:11	Jan 17 23:49	Jan 25 13:41
	Feb 01 05:46	Feb 08 13:50	Feb 16 16:57	Feb 23 22:32
	Mar 02 17:35	Mar 10 10:45	Mar 18 07:17	Mar 25 05:37
	Apr 01 06:24	Apr 09 06:47	Apr 16 18:55	Apr 23 11:56
	Apr 30 20:28 P	May 09 00:21	May 16 04:14 t	May 22 18:43
	May 30 11:30	Jun 07 14:48	Jun 14 11:52	Jun 21 03:11
	Jun 29 02:52	Jul 07 02:14	Jul 13 18:37	Jul 20 14:18
	Jul 28 17:55	Aug 05 11:06	Aug 12 01:36	Aug 19 04:36
	Aug 27 08:17	Sep 03 18:08	Sep 10 09:59	Sep 17 21:52
	Sep 25 21:54	Oct 03 00:14	Oct 09 20:55	Oct 17 17:15
	Oct 25 10:49 P	Nov 01 06:37	Nov 08 11:02 t	Nov 16 13:27
	Nov 23 22:57	Nov 30 14:36	Dec 08 04:08	Dec 16 08:56
	Dec 23 10:17	Dec 30 01:21		

All times are in Universal Time (UT1).
Eclipses at New Moon (solar eclipse) or Full Moon (lunar eclipse), are indicated by these symbols:

Solar Eclipses	Lunar Eclipses
T – Total	t – Total (Umbral)
A – Annular	p – Partial (Umbral)
H – Hybrid	n – Penumbral
P – Partial	

Phases of the Moon

Year	New Moon	First Quarter	Full Moon	Last Quarter
2023			Jan 06 23:08	Jan 15 02:10
	Jan 21 20:53	Jan 28 15:19	Feb 05 18:29	Feb 13 16:01
	Feb 20 07:06	Feb 27 08:06	Mar 07 12:40	Mar 15 02:08
	Mar 21 17:23	Mar 29 02:32	Apr 06 04:35	Apr 13 09:11
	Apr 20 04:12 H	Apr 27 21:20	May 05 17:34 n	May 12 14:28
	May 19 15:53	May 27 15:22	Jun 04 03:42	Jun 10 19:31
	Jun 18 04:37	Jun 26 07:50	Jul 03 11:39	Jul 10 01:48
	Jul 17 18:32	Jul 25 22:07	Aug 01 18:31	Aug 08 10:28
	Aug 16 09:38	Aug 24 09:57	Aug 31 01:35	Sep 06 22:21
	Sep 15 01:40	Sep 22 19:32	Sep 29 09:57	Oct 06 13:48
	Oct 14 17:55 A	Oct 22 03:29	Oct 28 20:24 p	Nov 05 08:37
	Nov 13 09:27	Nov 20 10:50	Nov 27 09:16	Dec 05 05:49
	Dec 12 23:32	Dec 19 18:39	Dec 27 00:33	

Year	New Moon	First Quarter	Full Moon	Last Quarter
2024				Jan 04 03:30
	Jan 11 11:57	Jan 18 03:53	Jan 25 17:54	Feb 02 23:18
	Feb 09 22:59	Feb 16 15:01	Feb 24 12:30	Mar 03 15:24
	Mar 10 09:00	Mar 17 04:11	Mar 25 07:00 n	Apr 02 03:15
	Apr 08 18:21 T	Apr 15 19:13	Apr 23 23:49	May 01 11:27
	May 08 03:22	May 15 11:48	May 23 13:53	May 30 17:13
	Jun 06 12:38	Jun 14 05:18	Jun 22 01:08	Jun 28 21:53
	Jul 05 22:57	Jul 13 22:49	Jul 21 10:17	Jul 28 02:51
	Aug 04 11:13	Aug 12 15:19	Aug 19 18:26	Aug 26 09:26
	Sep 03 01:55	Sep 11 06:06	Sep 18 02:34 p	Sep 24 18:50
	Oct 02 18:49 A	Oct 10 18:55	Oct 17 11:26	Oct 24 08:03
	Nov 01 12:47	Nov 09 05:56	Nov 15 21:29	Nov 23 01:28
	Dec 01 06:21	Dec 08 15:27	Dec 15 09:02	Dec 22 22:18
	Dec 30 22:27			

Year	New Moon	First Quarter	Full Moon	Last Quarter
2025		Jan 06 23:56	Jan 13 22:27	Jan 21 20:31
	Jan 29 12:36	Feb 05 08:02	Feb 12 13:53	Feb 20 17:33
	Feb 28 00:45	Mar 06 16:32	Mar 14 06:55 t	Mar 22 11:30
	Mar 29 10:58 P	Apr 05 02:15	Apr 13 00:22	Apr 21 01:36
	Apr 27 19:31	May 04 13:52	May 12 16:56	May 20 11:59
	May 27 03:02	Jun 03 03:41	Jun 11 07:44	Jun 18 19:19
	Jun 25 10:31	Jul 02 19:30	Jul 10 20:37	Jul 18 00:38
	Jul 24 19:11	Aug 01 12:41	Aug 09 07:55	Aug 16 05:12
	Aug 23 06:06	Aug 31 06:25	Sep 07 18:09 t	Sep 14 10:33
	Sep 21 19:54 P	Sep 29 23:54	Oct 07 03:47	Oct 13 18:13
	Oct 21 12:25	Oct 29 16:21	Nov 05 13:19	Nov 12 05:28
	Nov 20 06:47	Nov 28 06:59	Dec 04 23:14	Dec 11 20:52
	Dec 20 01:43	Dec 27 19:10		

All times are in Universal Time (UT1).
Eclipses at New Moon (solar eclipse) or Full Moon (lunar eclipse), are indicated by these symbols:

Solar Eclipses	Lunar Eclipses
T - Total	t - Total (Umbral)
A - Annular	p - Partial (Umbral)
H - Hybrid	n - Penumbral
P - Partial	

Phases of the Moon

Year	New Moon	First Quarter	Full Moon	Last Quarter
2026			Jan 03 10:03	Jan 10 15:48
	Jan 18 19:52	Jan 26 04:47	Feb 01 22:09	Feb 09 12:43
	Feb 17 12:01 A	Feb 24 12:28	Mar 03 11:38 t	Mar 11 09:39
	Mar 19 01:23	Mar 25 19:18	Apr 02 02:12	Apr 10 04:52
	Apr 17 11:52	Apr 24 02:32	May 01 17:23	May 09 21:10
	May 16 20:01	May 23 11:11	May 31 08:45	Jun 08 10:00
	Jun 15 02:54	Jun 21 21:55	Jun 29 23:57	Jul 07 19:29
	Jul 14 09:43	Jul 21 11:06	Jul 29 14:36	Aug 06 02:21
	Aug 12 17:37 T	Aug 20 02:46	Aug 28 04:18 p	Sep 04 07:51
	Sep 11 03:27	Sep 18 20:44	Sep 26 16:49	Oct 03 13:25
	Oct 10 15:50	Oct 18 16:13	Oct 26 04:12	Nov 01 20:28
	Nov 09 07:02	Nov 17 11:48	Nov 24 14:53	Dec 01 06:09
	Dec 09 00:52	Dec 17 05:43	Dec 24 01:28	Dec 30 18:59

Year	New Moon	First Quarter	Full Moon	Last Quarter
2027	Jan 07 20:24	Jan 15 20:34	Jan 22 12:17	Jan 29 10:55
	Feb 06 15:56 A	Feb 14 07:58	Feb 20 23:23 n	Feb 28 05:16
	Mar 08 09:29	Mar 15 16:25	Mar 22 10:44	Mar 30 00:54
	Apr 06 23:51	Apr 13 22:57	Apr 20 22:27	Apr 28 20:18
	May 06 10:58	May 13 04:44	May 20 10:59	May 28 13:58
	Jun 04 19:40	Jun 11 10:56	Jun 19 00:44	Jun 27 04:54
	Jul 04 03:02	Jul 10 18:39	Jul 18 15:45	Jul 26 16:55
	Aug 02 10:05 T	Aug 09 04:54	Aug 17 07:29 n	Aug 25 02:27
	Aug 31 17:41	Sep 07 18:31	Sep 15 23:04	Sep 23 10:20
	Sep 30 02:36	Oct 07 11:47	Oct 15 13:47	Oct 22 17:29
	Oct 29 13:36	Nov 06 08:00	Nov 14 03:26	Nov 21 00:48
	Nov 28 03:24	Dec 06 05:22	Dec 13 16:09	Dec 20 09:11
	Dec 27 20:12			

Year	New Moon	First Quarter	Full Moon	Last Quarter
2028		Jan 05 01:40	Jan 12 04:03 p	Jan 18 19:26
	Jan 26 15:12 A	Feb 03 19:10	Feb 10 15:04	Feb 17 08:08
	Feb 25 10:37	Mar 04 09:02	Mar 11 01:06	Mar 17 23:23
	Mar 26 04:31	Apr 02 19:15	Apr 09 10:27	Apr 16 16:37
	Apr 24 19:47	May 02 02:26	May 08 19:49	May 16 10:43
	May 24 08:16	May 31 07:37	Jun 07 06:09	Jun 15 04:27
	Jun 22 18:27	Jun 29 12:10	Jul 06 18:11 p	Jul 14 20:57
	Jul 22 03:02 T	Jul 28 17:40	Aug 05 08:10	Aug 13 11:45
	Aug 20 10:44	Aug 27 01:36	Sep 03 23:48	Sep 12 00:46
	Sep 18 18:24	Sep 25 13:10	Oct 03 16:25	Oct 11 11:57
	Oct 18 02:57	Oct 25 04:53	Nov 02 09:17	Nov 09 21:26
	Nov 16 13:18	Nov 24 00:15	Dec 02 01:40	Dec 09 05:39
	Dec 16 02:06	Dec 23 21:45	Dec 31 16:48 t	

All times are in Universal Time (UT1).
Eclipses at New Moon (solar eclipse) or Full Moon (lunar eclipse), are indicated by these symbols:

Solar Eclipses	Lunar Eclipses
T – Total	t – Total (Umbral)
A – Annular	p – Partial (Umbral)
H – Hybrid	n – Penumbral
P – Partial	

Phases of the Moon

Year	New Moon	First Quarter	Full Moon	Last Quarter
2029				Jan 07 13:26
	Jan 14 17:24 P	Jan 22 19:23	Jan 30 06:03	Feb 05 21:52
	Feb 13 10:31	Feb 21 15:10	Feb 28 17:10	Mar 07 07:52
	Mar 15 04:19	Mar 23 07:33	Mar 30 02:26	Apr 05 19:51
	Apr 13 21:40	Apr 21 19:50	Apr 28 10:37	May 05 09:48
	May 13 13:42	May 21 04:16	May 27 18:37	Jun 04 01:19
	Jun 12 03:51 P	Jun 19 09:54	Jun 26 03:22 t	Jul 03 17:58
	Jul 11 15:51 P	Jul 18 14:14	Jul 25 13:36	Aug 02 11:15
	Aug 10 01:56	Aug 16 18:55	Aug 24 01:51	Sep 01 04:33
	Sep 08 10:44	Sep 15 01:29	Sep 22 16:29	Sep 30 20:57
	Oct 07 19:14	Oct 14 11:09	Oct 22 09:28	Oct 30 11:32
	Nov 06 04:24	Nov 13 00:35	Nov 21 04:03	Nov 28 23:48
	Dec 05 14:52 P	Dec 12 17:49	Dec 20 22:46 t	Dec 28 09:49

Year	New Moon	First Quarter	Full Moon	Last Quarter
2030	Jan 04 02:49	Jan 11 14:06	Jan 19 15:54	Jan 26 18:14
	Feb 02 16:07	Feb 10 11:49	Feb 18 06:20	Feb 25 01:58
	Mar 04 06:35	Mar 12 08:48	Mar 19 17:56	Mar 26 09:51
	Apr 02 22:02	Apr 11 02:57	Apr 18 03:20	Apr 24 18:39
	May 02 14:12	May 10 17:11	May 17 11:19	May 24 04:57
	Jun 01 06:21 A	Jun 09 03:36	Jun 15 18:41 p	Jun 22 17:19
	Jun 30 21:34	Jul 08 11:02	Jul 15 02:12	Jul 22 08:07
	Jul 30 11:11	Aug 06 16:43	Aug 13 10:44	Aug 21 01:15
	Aug 28 23:07	Sep 04 21:55	Sep 11 21:18	Sep 19 19:56
	Sep 27 09:55	Oct 04 03:56	Oct 11 10:47	Oct 19 14:50
	Oct 26 20:17	Nov 02 11:56	Nov 10 03:30	Nov 18 08:32
	Nov 25 06:46 T	Dec 01 22:57	Dec 09 22:40 n	Dec 18 00:01
	Dec 24 17:32	Dec 31 13:36		

All times are in Universal Time (UT1).
Eclipses at New Moon (solar eclipse) or Full Moon (lunar eclipse), are indicated by these symbols:

Solar Eclipses	Lunar Eclipses
T - Total	t - Total (Umbral)
A - Annular	p - Partial (Umbral)
H - Hybrid	n - Penumbral
P - Partial	

Photo 3–2 The changing phases of the Moon. ©2010 F. Espenak

The phases of the Moon for the entire 21st Century can be found at the AstroPixels.com website:

www.astropixels.com/ephemeris/moon/phases2001gmt.html

ECLIPSE ALMANAC: 2021 TO 2030

Books by the Author

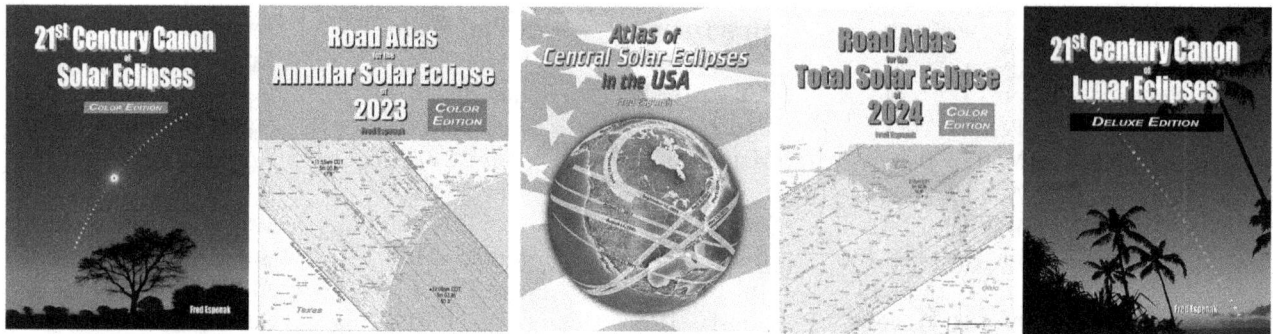

Below is a partial list of some of the books Fred Espenak has written through Astropixels Publishing:

21st Century Canon of Solar Eclipses

The complete guide to every solar eclipse occurring from 2001 tom 2100 (224 eclipses in all). It includes information and maps for all total, annular, hybrid, and partial eclipses. A special world atlas shows detailed full page maps of all central eclipse paths (total, annular and hybrid).

Road Atlas for the Annular Solar Eclipse of 2023

Detailed road maps of the entire eclipse path from the western USA, through Mexico, Central and South America. Information printed on the maps makes it easy to estimate the duration of annularity from any location in the eclipse path. This is the next annular solar eclipse visible from the USA.

Atlas of Central Solar Eclipses in the USA

When was the last total eclipse through the USA and when is the next? How often do they happen? What total eclipse tracks passed across the USA during the 17th, 18th, and 19th centuries, etc., and what states did they include? And how often is a total solar eclipse visible from each of the 50 states? The Atlas of Central Solar Eclipses in the USA answers all of these questions and more with hundreds of maps and tables.

Road Atlas for the Total Solar Eclipse of 2024

This book contains detailed road maps of the entire eclipse path from Mexico, through the USA and Canada. Information printed on the maps makes it easy to estimate the duration of totality from any location in the eclipse path. This is the next total solar eclipse visible from the USA.

21st Century Canon of Lunar Eclipses

The complete guide to every lunar eclipse occurring from 2001 tom 2100 (228 eclipses in all). It includes information and maps for all total, partial, and penumbral eclipses. The predictions use a new model for Earth's elliptical shadows.

For information on these books and more, visit *Astropixels Publishing*:

http://eclipsewise.com/pubs/index.html

www.ingramcontent.com/pod-product-compliance
Lightning Source LLC
Chambersburg PA
CBHW080056280326
41934CB00014B/3338